AF572488

TROPICAL VEGETABLE PRODUCTION

TROPICAL VEGETABLE PRODUCTION

By
G Jahir Hussain

2012

SBS Publishers & Distributors Pvt. Ltd.
New Delhi

All rights reserved. No part of this publication may be reproduced, stored in a retrieval system, or transmitted in any form or by any means, electronic, mechanical, photocopying, recording or otherwise, without the prior written permission of the publisher and the copyright holder.

ISBN 13 : 9789380090481

First Published in 2012

© Reserved

Published by:

SBS PUBLISHERS & DISTRIBUTORS PVT. LTD.
2/9, Ground Floor, Ansari Road, Darya Ganj,
New Delhi - 110002,
INDIA
Tel: 0091.11.23289119 / 41563911 / 32945311
Email: mail@sbspublishers.com
www.sbspublishers.com

Printed in India by Chaman Enterprises, New Delhi.

Content

Preface

Sustainable crop production is vital to ensure that supplies of fresh vegetables and their products are readily available. The food security still remains a huge problem in areas of the world, including the tropics and sub-tropics, where communities rely solely on subsistence farming to meet their day to day food demands. It is evident that food production needs to become more sustainable to ensure economic stability and poverty reduction. With this in mind Tropical Vegetable Production addresses the problems surrounding vegetable production in developing countries. It has great role in describing the principles and practise of tropical vegetable production, from site selection, security and management to seeds, crop preparation and pesticides, and those crops which are of particular importance in developing countries.

Author

1

Fats and Oils

NATURE AND SOURCES OF FATS AND OILS

CHEMICAL AND PHYSICAL CHARACTERISTICS

The present study deals only with such oils and fats as are capable of serving as foodstuffs, even though in practice they are not put to such use. The common chemical characteristic of such oils and fats is that they may be decomposed into glycerin and one or more acids of the class known to chemists as fatty acids. Chemists designate as acids a class of substances which have an acid or sour taste; contain the element hydrogen; and act upon metals, hydrogen being evolved and its place being taken by the metal.

The compound thus formed with the metal is known as a salt. The common physical properties of such oils and fats are that they float on water but are not soluble in it; they are greasy to the touch, and have lubricating properties; they are not readily volatile; and may be burned without leaving any residue, *i.e.*, ash. No other class of substances has the chemical properties of the fats and oils; but many possess similar physical ones, *e.g.*, mineral oils, earth-wax (ozocerite), paraffin, animal waxes like spermaceti or beeswax, vegetable waxes like carnauba or candelilla wax, volatile or essential vegetable oils like the oils of thyme, of cloves, of cedar, and attar of roses. None of these substances furnishes both glycerin and fatty acids; none of them has nutritive value; none of them will be considered further in this treatise. Fats and oils, then, in the restricted meaning in which these two words are used hereafter, are substances which consist always of chemical combinations of glycerin with certain fatty acids, and which may serve as foods.

The distinction between a fat and an oil is purely an accidental one depending upon the environment in which the substance happens to be placed. If the substance is solid at ordinary temperatures, it is termed a fat; if fluid, an oil. This is merely a distinction of convenience, since all oils are solidified at lower temperatures and all fats melted at higher temperatures.

Obviously, the dividing line that holds for a cool climate would not hold for a hot one. In each climate, however, the distinction is of importance in industrial and in culinary uses; it has also some importance in nutrition, since fats are somewhat less digestible than oils. In this study *fat* is often used indiscriminately for a solid or a liquid substance of the class here under consideration.

ANIMAL AND VEGETABLE SOURCES

Animal fats and oils are derived both from terrestrial and marine animals. Marine fats include liver oils, blubber oils, and fish oils. In addition, from certain marine animals waxes are obtained, *e.g.*, spermaceti, which, because it is a wax and not a fat, need not be considered here.

The different types of marine fats, which in practice are often mixed, have been of great importance in the past and still possess considerable significance. Some of these serve special purposes, such as codliver oil; others are used to some extent as foodstuffs; but for the most part they serve industrial uses. With two important exceptions animal fats are obtained from carcasses. These two exceptions are butter and the fat of the yolks of eggs. Carcass fat is found in different locations. There is a good deal of it in the visceral cavities and in and around the viscera. More or less of it occurs in the muscles, in the connective tissue, under the skin, and in the bones. The proportions found in the different parts of the body vary from species to species and in any given species with the age of the individual animal and its condition. The fats from the different parts and organs of a given animal differ somewhat in their properties. As a rule, the fat from the interior of the animal is somewhat firmer than the fat from near the body surface, *i.e.*, it melts at a somewhat higher temperature. Furthermore, under certain conditions the feed of the animal affects the physical properties of the carcass fat more or less. Animals fattened upon a diet containing much oil—for example, peanuts—tend to produce softer carcass fats than animals of the same species fattened upon a diet containing relatively little oil—for example, corn (maize). Animals form fairly specific fats from starches but deposit in their tissues unchanged such portion of the fats and oils of the ration as is not promptly oxidized.

Vegetable fats and oils are found in greatest abundance in fruits and seeds. While fats and oils do occur in the roots, stalks, branches, and leaves of plants, they are rarely present in these organs in quantities large enough for commercial purposes. In some seeds and fruits, however, the fat content is great—in several cases as high as 35 per cent; in dried coconuts 65 per cent—and these are the commercial sources of vegetable fats. In some seeds the fat is practically confined to the germ or embryo; this is the fact in most of the cereals.

(The germ or embryo is that part of the seed which gives rise to the plant when the seed germinates. It is usually only a small part of the seed. The remainder of the seed consists mainly of reserve food material with the help of which the embryo grows into the plantlet which, as it develops roots, is enabled to draw its nourishment from the soil.) The olive contains a large amount of fat in the pulp surrounding the kernel and only a smaller amount in the kernel itself, while in the oil palm both the pulp and the kernel contain large amounts. The fat from the pulp may have characteristics quite different from those of the fat in the kernel.

EDIBLE AND INEDIBLE FATS

In commerce a distinction is commonly made between *edible*

and *inedible* fats, based either upon external characteristics, such as unattractive colour, taste, or odour, or upon sentimental considerations, such as revolting origin (from garbage, for example), decomposition, or the possibility of contamination with a poisonous substance or with the germs of disease. The distinction between edible and inedible fats is nevertheless a purely practical one, for with modern methods nearly all fats can be refined or modified to the point of physiological edibility. That the distinction exists at all is because it is either unprofitable to convert inedible into edible fat to a greater extent than is done or else because such conversion is not permitted for sanitary reasons.

Sanitary considerations are a more important factor deterring the transformation of inedible into edible fats in the case of animal than in the case of vegetable products, for animal fats may be treated as inedible, even if they are not repulsive to the senses, because their origin is revolting. This is the case when they are obtained from animals that have died otherwise than by slaughter, *i.e.*, from disease, old age, or accident. Such fats are not permitted by health authorities to be used for food purposes because of the danger of transmitting disease, though in former times some fat of this sort was unquestionahly so used.

Because of the danger of disease transmission, fat from animals killed by slaughter is not permitted to be used for food if inspection of the carcass shows that the animals were diseased. In most countries inspections have been established in slaughterhouses to protect the consumer from this danger. The degree and effectiveness of such inspections vary in different sections and in different countries. In the United States, food fats remain subject to the provisions of federal and local food laws after they are shipped out of the inspected slaughterhouse. They may, therefore, if officials deem it necessary, again be inspected at any time on their way to the consumer.

So far as is known to date, there is little or no danger of the transmission of diseases of plants to man. Therefore, governments have not thought it necessary to supervise the production of vegetable fats in the same rigid manner in which fat production from animal carcasses is controlled. The only sanitary control over vegetable oils that exists, aside from the very special control applied to the manufacture of margarin, is the general control exercised by the several food laws over vegetable oils in common with all other foodstuffs. This form of control concerns itself principally with adulteration, with decomposition (rancidity), and with various types of misbranding. Vegetable oils, in common with nearly all other foodstuffs, may transmit disease if handled by infected persons. Such occurrences are extremely rare and of a type not at present controllable by food officials.

DRYING AND NON-DRYING OILS

In commerce the distinction between edible and inedible fats is not the only one that is made. A different but equally important distinction is drawn between the drying and the non-drying oils. The two kinds of distinction are not comparable, for both non-drying and drying oils may be either edible or inedible. Thus in Europe linseed oil (a typical drying oil) is used for food, whereas in America it is not now so used—for one reason because it is too expensive as compared with non-drying oils which are readily available in abundance.

Drying oils absorb oxygen from the air and are thereby converted into plastic, elastic, resin-like substances. Hence, when exposed in a thin layer, as in painting, they form a tough, elastic, waterproof film which adheres tightly to the painted surface and protects it from the weather. The two chief drying oils are linseed oil and tung (chinawood) oil, which find wide use in the manufacture of paints, varnishes, artificial rubber, linoleums, and other coverings.

The non-drying oils find a wide variety of industrial uses: they enter into soaps and cleansers, cosmetics, lubricants, leather dressings, and candles. They are used in the processes of wool manufacture, especially carding; they are employed in making tin plate and in foundry work. Fats and oils, whether edible or inedible, drying or non-drying, have still other industrial uses. They were the primitive illuminants and are still so used in a relatively slight degree. Some of the industrial uses of the fats and oils depend upon their physical properties, others upon their chemical character, to which detailed reference will be made later.

Between the drying oils and the non-drying oils is a group of oils which, while they possess the property of absorbing oxygen, do not do so sufficiently to qualify them as drying oils. They are changed more or less when exposed to the atmosphere, but not as completely as linseed, tung, and certain other drying oils.

They are often termed semi-drying oils. All gradations are found between completely drying oils and completely non-drying oils. Soy bean oil and corn oil are examples of semi-drying oils. When exposed to the atmosphere in a thin layer they thicken but do not form a hard, dry film. The film remains sticky and somewhat runny—tacky is the word often used. Under some circumstances appropriate amounts of semi-drying oils are mixed with full-drying oils to make paints of cheaper grades.

DETERIORATION OF FATS AND OILS

Fats and oils are quite unstable substances. When stored for any considerable length of time, especially when the temperature is high and the air has free access to them, they deteriorate and spoil. In this respect different fats differ markedly. Some spoil very much more rapidly than others. Among the various fats, spoilage takes the form of rancidity. The fat acquires a peculiarly disagreeable odour and flavour. A vast amount of scientific research has been carried on to determine the cause and nature of rancidity, but investigators are far from agreement on the subject. For present purposes it is sufficient to point out that spoilage of a fat, usually identical with rancidity, is accompanied by partial splitting of the fat into glycerin and fatty acids. The glycerin disappears, or at any rate is unobjectionable, but the fatty acids remain dissolved in the fat, give it an acid reaction, and contribute to its objectionable rancid flavour.

The rancidity of a given parcel of fat is not necessarily the result of long storage under unfavourable conditions. The fat may have been spoiled and rancid from the moment of its production. This will inevitably be true when the materials from which it was produced have undergone decomposition. Thus the fat obtained from putrefying carcasses will be rancid and so will the oil expressed from fermented cottonseed.

In other words, to obtain a sound and sweet fat, the raw material must be sound and sweet; it must be worked up speedily before it has had time to decompose; and this must be done under clean and sanitary conditions. The fat thus obtained must be stored under favourable conditions and its consumption cannot be too long delayed.

These conditions it is difficult to obtain in many of the less civilized portions of the world, especially in the tropics, where many fat- and oil-yielding raw materials are produced. Hence fats and oils made at the source of the raw materials may be less sound than those produced at or near the place of consumption.

The fact that so great a proportion of the fat supply, especially vegetable oils, is or becomes rancid and decomposed, necessitates refining, decolorizing, and deodorizing. Certain oils —for example, cottonseed oil—require refining even when they are not decomposed, because they contain certain impurities and are of dark colour. The act of refining is not merely an item of expense, but in the case of decomposed fats it involves the removal of the fatty acid contaminating the fat or oil. The effect is that the yield of refined oil is less than the crude oil with which the operation was begun.

This disappearance of material is known in the trade as the refining loss. Hence the trade in fats, oils, and greases specifies the amount of free fatty acids permissible in them. If the amount exceeds specifications, price adjustments are commonly demanded and conceded. If the refining loss is too heavy to make refining economical, the fat is commonly consigned to the soap kettle. The past history of a fat —whether or not it was ever decomposed —in some way not clearly understood, affects its keeping quality after refining. Such fats require especially great care and skill in refining, deodorizing, and decolorizing to insure reasonable keeping qualities. It may not be profitable to refine a fat even though the refining loss be only moderate.

PROPERTIES OF FATS AND OILS

For an understanding of the place of fats and oils in the diet and in the arts, some elementary knowledge of their chemical and physical properties is essential. It is the object of this section to present the minimum of such necessary information in the simplest way practicable.

CHEMICAL COMPOSITION

As already stated, fats may be decomposed into glycerin and fatty acids. This manner of decomposition takes place only in the presence of moisture. For each molecule (a molecule is the smallest particle of a substance that can exist and still exhibit the properties of that substance) of glycerin set free there are set free three molecules of fatty acid.

In the process three molecules of water are taken up, partly to help re-form the glycerin and partly to help re-form the fatty acids. Conversely (in the laboratory) the fat may be reconstituted from glycerin and fatty acid, in which event three molecules of water are set free for each molecule of fat synthesized.

The process of splitting a substance whereby water is taken up is known to chemists as hydrolysis, a word which is merely Greek for cleavage by water. The process is often termed saponification, since it was first observed to take place in the manufacture of soap. The term saponification (instead of the more exact term hydrolysis) is, however, applied indiscriminately and inappropriately to any chemical change of this nature, whether or not soap is formed. Nowadays in industry fats are very often converted into glycerin and fatty acids—that is, hydrolyzed—without the formation of any soap whatever.

A soap is merely the combination of a fatty acid with a metal, *i.e.,* it is a salt. The commonest soaps are the fatty-acid salts of sodium (sodium is a soft, white metal obtained from common salt, sodium chloride) and potassium. (Potassium is also a soft, white metal obtained from wood ashes or from certain minerals found in Germany, Alsace, and elsewhere. Both sodium and potassium oxidize with great rapidity when exposed to the air, and hence are never found in nature except in the form of their compounds.)

Hard soaps are sodium salts; soft soaps, potassium salts. The fatty-acid salts of ammonium are also sometimes used for cleansing. Only a few other soaps are of practical importance, for example lead soaps which are used in medicinal plasters, zinc soaps which are used in ointments, and aluminum soaps which are used in waterproofing.

Very few of the salts of fatty acids have the properties of common soap. Most of them are but slightly soluble in water, and therefore do not yield suds and have little or no detergent (*i.e.*, cleansing) action. All are nevertheless termed soaps by chemists.

TRIGLYCERIDS AND FATTY ACIDS

As above stated, fats may be split into glycerin and fatty acids, the resulting mixture containing three molecules of fatty acid for each molecule of glycerin. Because of this proportion of acid to glycerin, the chemical compounds found in the fat before it was split are known to chemists as triglycerids. Since there are a number of different fatty acids that occur in natural fats, a great many different triglycerids are encountered in nature. These are named according to the fatty acid or acids they contain.

Thus triolein is the triglycerid of oleic acid, tripalmitin that of palmitic acid, tristearin that of stearic acid, while monopalmitin-distearin contains, as the name indicates, one molecule of palmitic and two of stearic acid. While a large variety of fatty acids is found in natural fats and oils, only a few of them are of outstanding commercial importance. These are myristic acid, lauric acid, palmitic acid, stearic acid, oleic acid, linolic acid, and linolenic acid. Though the number of triglycerids encountered in nature is great, the triglycerids of these seven acids make up the great bulk of the natural fats and oils. Fats and oils are practically always mixtures of triglycerids in varying proportions.

In some fats one triglycerid predominates, in others another, and in still others several are present in material amounts. Apparently no natural fat or oil consists solely of a single triglycerid. The properties of different fats and oils depend upon the characteristics of the triglycerids of which they are mixtures and upon the proportions of these triglycerids to one another.

The fats of different species of animals and plants vary widely. Indeed, the fat from a given natural source, say a given species of animal or plant, may contain the same triglycerids in slightly different proportions, depending upon the conditions of the environment prevailing while the fat was being formed. It was also pointed out that a fruit may yield two fats of different properties, one from the pulp and one from the kernel. In the case of plants the fat may also vary with the cultural variety of the plant and with the climatic and soil conditions under which the plant was grown.

Thus the linseed oils from Argentina, India, Russia, and the United States have slightly different chemical and physical properties.

The formulas of these acids (disregarding isomers) are as follows:

Acid	Elementary Formula	Constitutional Formula
Lauric	$C_{12}H_{24}O_2$	$CH_3(CH_2)_{10}COOH$
Myristic	$C_{14}H_{28}O_2$	$CH_3(CH_2)_{12}COOH$
Palmitic	$C_{16}H_{32}O_2$	$CH_3(CH_2)_{14}COOH$
Stearic	$C_{18}H_{36}O_2$	$CH_3(CH_2)_{16}COOH$
Oleic	$C_{18}H_{34}O_2$	$CH_3(CH_2)_{14}(CH)_2COOH$
Linolic	$C_{18}H_{32}O_2$	$CH_3(CH_2)_{12}(CH)_4COOH$
Linolenic	$C_{18}H_{30}O_2$	$CH_3(CH_2)_{10}(CH)_6COOH$

Investigators may find somewhat different proportions, but in general these are representative:

Fat or oil	Lauric	Myristic	Palmitic	Stearic	Oleic	Linolic	Linolenic
Coconut	45	20	5	3	6	–	–
Palm kernel	55	12	6	4	10	–	–
Tallow (beef)	–	2	29.0	24.5	44.5	–	–
Tallow (mutton)	–	2	27.2	25.0	43.1	2.7	–
Lard	–	–	24.6	15.0	50.4	10.0	–
Olive	–	–	14.6	–	75.4	10.0	–
Arachis (peanut)	–	–	8.5	6.00	51.6	26.0	—
Cottonseed	–	–	23.4	–	31.6	45.0	–
Maize	–	–	6.0	2.0	44.0	48.0	–
Linseed	–	3	6.0	–	–	74.0	17.0
Soy bean	–	–	11.0	2.0	20.0	64.0	3.0

Fats and oils being mere mechanical mixtures of triglycerids, it is possible in many cases to separate them more or less completely into their component triglycerids by simple mechanical means, chilling and pressure. Such processes have considerable commercial importance, as, for example, the separation of lard into lard oil and lard stearin or of beef tallow into oleo oil and oleostearin.

For the details of the chemical nature of the fatty acids, the reader is referred to the texts on organic chemistry. Here it is sufficient to point out that they all possess the characteristic property of acids in general, *viz.*, to combine with bases to form salts. These salts, as was pointed out above, are known as soaps whether or not they have detergent action. Moreover, all fatty acids contain carbon, hydrogen, and a small proportion of oxygen. They differ from one another in the number of carbon atoms in each molecule, in the proportion of carbon to oxygen their molecules contain, and also in the proportion of carbon to hydrogen. Upon these ratios the physical and chemical properties of the acids and of their triglycerids very largely depend.

SATURATED AND UNSATURATED FATTY ACIDS

For present purposes it is sufficient to limit consideration to one of the aspects of the carbon-hydrogen ratio of the fatty acids. When the fatty-acid molecule contains the maximum of hydrogen possible, the acid is said to be a saturated fatty acid. It is saturated with respect to hydrogen. Myristic, lauric, palmitic, and stearic acids are such saturated acids. They are solids at ordinary temperatures. When, however, the fatty-acid molecule does not contain the maximum amount of hydrogen possible, the acid is said to be an unsaturated fatty acid.

It is unsaturated with respect to hydrogen. Such unsaturated acids are oleic, linolic, and linolenic acids. They are liquids at ordinary temperatures. By chemical means these acids may be made to take up, *i.e.,* combine with, hydrogen. This process is known as hydrogenation. It converts a more unsaturated fatty acid into a less unsaturated one, or, if the hydrogenation is carried to completion, into a saturated fatty acid. Thus by hydrogenation oleic acid is converted into stearic acid. Linolic acid when hydrogenated can be made to take up twice as much hydrogen as oleic acid, and linolenic acid three times as much. Linolic acid is, therefore, a more highly unsaturated acid than oleic, while linolenic acid is more highly unsaturated than linolic.

Unsaturated fatty acids can be made to combine with other substances instead of with hydrogen. For example, they may be made to take up iodin or oxygen. Acids of a low degree of unsaturation, such as oleic acid, do not combine with oxygen with any great degree of avidity, but acids of a greater degree of unsaturation, such as linolic or linolenic, combine with it very readily; they do so merely upon exposure to the air.

The properties of the fatty acids just described, which depend upon their degree of saturation or unsaturation with respect to hydrogen, are retained by them when they are in combination with glycerin as triglycerids. Hence the different fats are also more or less saturated, according as they contain greater or lesser proportions of the triglycerids of saturated or unsaturated fatty acids. When fats contain large amounts of trilinolin and trilinolenin, these absorb oxygen avidly. It is upon the presence, then, of these unsaturated triglycerids that the properties of drying oils described in the preceding chapter depend. Resin-like films are formed by them when oxygen is absorbed because the oxidation products of these unsaturated triglycerids are relatively insoluble solids. Upon this reaction, as already pointed out, is based the behaviour of paints.

MEASURES OF UNSATURATION

It is obvious, then, that it is important for industrial users of fats to know the degree of unsaturation of a given parcel of fat. This might be ascertained by determining the amount of hydrogen required to convert it into a saturated fat. In practice this is a complicated procedure and so simpler methods are resorted to. The simplest of these is the determination of the amount of iodin that can be made to combine with the fat. The percentage by weight of iodin absorbed by the fat in the natural state is known as the *iodin number.* It is an index to the degree of unsaturation of the fat. Examination of the table shows that the fats with the highest iodin numbers are the drying oils par excellence, linseed and tung oil, with which must also be classified menhaden fish oil.

Table. Iodin Numbers of Common Fats

Fat or Oil	Lodin number
Linseed oil	173 – 201
Tung Oil	170.6
Menhaden oil	139 – 173
Whale oil	121 – 146.6
Soy bean oil	137 – 143
Sunflower oil	119 – 135
Corn oil	111 – 130
Cottonseed oil	108 – 110
Sesame oil	103 – 108
Rapeseed oil	94 – 102
Peanut oil (arachis)	83 – 100
Olive oil	79 – 88

Horse oil	71 – 86
Lard	46 – 70
Palm oil	51.5 – 57
Milk fat	26 – 50
Beef tallow	38 – 46
Mutton tallow	35 – 46
Cacao butter	32 – 41
Palm kernel oil	13 – 17
Coconut oil	8 – 10

OTHER USEFUL TESTS

There are, of course, many other tests besides iodin absorption that are used in commercial practice. This is not the place to discuss them in detail. A few of them, however, deserve mention in passing.

The iodin number of a fat tells us the degree of unsaturation of a fat. It does not tell us whether the unsaturation is the result of the presence of triolein only, of trilinolin only, of trilinolenin only, or of a mixture of the three. As the drying qualities depend mainly upon trilinolin and trilinolenin, paint manufacturers are not always satisfied with the determination of the iodin number. In such cases they determine the amount of oxygen the oil tested will absorb under standard conditions. As the absorption of oxygen is mainly by the trilinolin and trilinolenin, this test is used to supplement the iodin number.

The amount of free fatty acid is estimated by determining the quantity of alkali that must be added to the fat to render it quite neutral. Sometimes, in addition to estimating the free fatty acid in this way, the actual loss in refining is also determined. This is done by warming a known amount of the fat with strong aqueous caustic soda solution, which converts the free fatty acid into soap. (Caustic soda is a compound of one atom each of sodium, oxygen, and hydrogen; its formula is therefore NaOH. Its proper scientific name is sodium hydroxid. It is also known as soda lye or simply as lye. It is very alkalin and corrosive.) This soap is then removed and the amount of fat remaining is then determined. The loss is estimated by subtracting this amount from the amount of fat originally taken for the test. The amount and strength of caustic soda solution, the temperature, and the length of treatment are so chosen that only the free fatty acid and other impurities present in the oil are removed and but little, if any, saponification of neutral fat takes place.

Many fats and oils contain substances that are not triglycerids. These may be natural constituents or they may be adulterants or contaminants. The presence of a considerable proportion of them of course reduces the commercial value of the fat. The commonest of these is moisture. It is estimated very simply by placing a weighed portion of the fat in an oven heated to a temperature slightly higher than that of boiling water. The moisture is thereby driven off. The fat is then again weighed; the loss is regarded as moisture.

The determination of non-fat materials other than water is done by saponifying the fat by heating with strong caustic soda or potash solution until all the triglycerids have been decomposed into glycerin and soap. (Caustic potash is the compound of potassium analogous to caustic soda.) These are soluble in water and may be washed away. What remains behind is the non-triglycerid part of the fat and may be weighed. It is known as the unsaponifiable matter. In practice the procedure is not as simple as this, but the basic principle is correctly stated above.

The determination of unsaponifiable matter must not be confused with the saponification number of a fat. The saponification number is the number of milligrams of potassium hydroxid required to convert one gram of the fat completely into glycerin and potassium soap. It gives information concerning the character of the fatty acids of the fat and in particular concerning the solubility of their soaps in water. The higher the saponification number of a fat free from moisture and unsaponifiable matter, the more soluble the soap that can be made from it.

These acids occur in undecomposed butter in chemical combination as triglycerids. Their sodium soaps are quite soluble in water. The high saponification number of coconut oil and palm kernel oil is due to the large proportion of lauric acid and myristic acid that they contain. These oils therefore yield quite soluble soaps.

Table. Saponification Numbers of Common Fats

Fat or Oil	Saponification Number
Rapeseed oil	170 - 179
Menhaden oil	190.6
Corn oil	188 - 193
Olive oil	185 - 196
Soy bean oil	193
Cacao butter	193.55
Linseed oil	192 - 195
Cottonseed oil	193 - 195
Lard	195.4
Mutton tallow	192 - 195.5
Peanut oil (arachis)	190 - 196
Horse oil	195 - 197
Beef tallow	193.2 - 200
Palm oil	196 - 205
Butter	220 - 233
Palm kernel oil	242 - 250
Coconut oil	246 - 260

Before leaving the subject of the commercial chemical testing of fats, the *titre test* deserves mention because it is of much importance in certain branches of industry. The *titre* of a fat or oil is the temperature at which the mixture of fatty acids derived from it solidifies after it has been melted. The test is performed in several steps. First, the fat is completely saponified, usually by heating with a solution of caustic soda. Then the mixture of soaps thus obtained is treated with a strong acid, usually sulfuric, which takes the sodium away from the soaps, thereby converting them into free fatty acids. After these have been washed and dried they are melted and the temperature at which the melted mass solidifies when cooled is noted. This temperature gives an index to the consistency of the original fat, a matter of great importance to manufacturers of candles and of products like margarin, in which consistency and texture are of the utmost importance.

Finally, the viscosity of a fat is a property of commercial significance, especially to manufacturers of lubricants. It is usually estimated by comparing the length of time it takes a given volume of oil (or melted fat) to flow through a tube of small bore, or through a small orifice, with the time it takes an identical volume of water. Castor oil has the highest viscosity of any fat that is fluid at ordinary temperatures.

FATS AND OILS TECHNOLOGY

The economic aspects of the fat and oil trade are so interwoven with the uses of these raw materials that some elementary consideration of the technology of fats and oils is necessary for a proper understanding of the economic situation.

COMMERCIAL PRODUCTION OF ANIMAL FATS

Animal fats—except butter—are separated by a process termed rendering or trying-out. The simplest method is to throw the fat-containing tissues into a kettle, heat them over an open fire till the fat has been cooked out, and then strain off the fat from the non-fatty material which is known as cracklings or greaves. More commonly the kettle is not heated by a direct flame but by steam, to avoid charring. This is done in a number of ways. The simplest is with a steam coil which projects into the interior of the kettle and comes in contact with the material to be rendered.

Another common method is to use a *jacketed* kettle. This has a double wall and the heating is done by passing steam through the space between the walls. Sometimes, however, fats are not rendered dry but are placed in the kettle with water and then heated in any of the ways mentioned above; or else steam is blown directly into the water, a procedure wasteful of fuel. The commonest rendering method of all is with steam under pressure at a temperature of 120-130 deg Centigrade. The material to be rendered is placed in a large strong metal cylinder known as a digester, which is then closed tightly. Steam under pressure is passed into it for some time—often as long as twelve to fifteen hours. The steam is then shut off, the apparatus is allowed to cool, and the fat is skimmed off the surface of the water, some of which was placed in the digester with the material to be rendered and some of which condensed from the steam during the operation. This is an efficient method of rendering, though it does not yield by any means the best product. It is very economical in regard to consumption of fuel and at these high temperatures tissues, even bones, release their fat quite completely. In consequence, the yields by this method are greater than by the others.

When the process of rendering is such as to leave cracklings (scrap) or greaves, these retain considerable fat, most of which is often recovered by some form of expression either with a hydraulic press or an expeller.

The best quality of animal fat is obtained by rendering perfectly fresh material with water at low temperatures. It is in this manner that neutral lard and oleo stock are prepared for the manufacture of oleomargarin.

HOG FATS OR LARD

The best grade of lard is known as neutral lard No. 1. It is obtained from the leaf fat of the hog, mainly kidney fat, and fat in the omentum. The omentum is a thin sheet of tissue attached to the intestines; in well-nourished animals it contains a good deal of fat. Before the development of the margarin industry, that is, before 1875, in great part it was rendered separately and marketed as leaf lard, commanding a higher price than other lard because of its great firmness and better flavour. It is still so marketed to some extent by smaller butchers and the small packing houses. In the larger establishments the omental and kidney fat are removed from the carcass at the earliest feasible moment, cooled and promptly rendered with pure water at a low temperature, usually not higher than 50 deg Centigrade.

The product is known as neutral lard No. 1 and is used almost exclusively in the manufacture of oleomargarin.

This process does not recover all of the lard in the parts rendered. The remainder is recovered by cooking in digesters at higher temperatures and pressures. It is sold as leaf lard. It should be noted that it is a somewhat different product from the leaf lard of former times and from the leaf lard produced by small concerns that do not make neutral lard, for it contains only a portion of the leaf and kidney fat instead of all of it.

The manner of cooling melted lard greatly affects its appearance. If allowed to cool without special precautions, it is apt to be translucent and to have the appearance which the American consumer attributes to a grease rather than the white, opaque, somewhat granular appearance which he expects in lard. It is therefore customary in packing houses to chill it rapidly. This is usually done with so-called lard rolls. These are large, smooth, hollow, metal cylinders which are revolved on a horizontal shaft.

They are cooled from the inside with a current of brine, and the lard is run onto one side of their surface in a thin layer, thereby being chilled quickly. It is then scraped off automatically by a stationary scraper on the other side. The chilled lard drops into a trough in which is a worm conveyor or *picker* which churns up the lard, thereby giving it the desired colour and texture, and conveying it still in the plastic state into storage tanks from which it flows into the containers for shipment. Sometimes the beating up of the plastic lard is so done as to incorporate air which makes it appear whiter and increases its volume though not its weight.

GREASES

Grease, which in packing-house parlance is merely inedible lard, is rendered with steam under pressure from packing-house offal, and from carcasses that have been condemned by the governmental meat inspectors as being unfit for food for one reason or another. It is graded according to colour as white, yellow, or brown grease. White and yellow packing-house greases commonly contain only hog fat. Brown grease may also contain some beef and mutton fat, for it includes the fat from the catch basins. In large packing houses great quantities of water are used which ultimately find their way into the sewers, carrying along in their passage not inappreciable quantities of fats.

To recover these it is customary to let the waters settle in basins before they finally flow into the sewer. In these basins fat rises to the surface; it is then skimmed off and combined with the brown grease. As the waters come from all parts of the packing house in which cattle and sheep may be slaughtered as well as hogs, the brown grease naturally may contain beef and mutton fat.

Grease is also recovered from certain of the viscera, other offal, and especially the intestines. In former times, before the practice was restricted by the government, this was known as gut lard and mixed with other lard for edible purposes. It has a characteristic flavour which experts are able to recognize even in admixture. In slaughterhouses under federal inspection all offal of this kind must be worked up for grease.

Greases are sold on the basis of their colour and of their chemical composition. The percentage of free fatty acid, of unsaponifiable matter, and of moisture are taken into consideration, and also the titre test. Greases are used chiefly in the manufacture of soap, candles, and lubricants. Before the enactment of the meat-inspection act in 1906 undoubtedly much that is now sold as grease was used for edible purposes. Today the meat-inspection act requires that such greases be denatured by the addition of petroleum products or other denaturant.

CATTLE AND SHEEP FATS OR TALLOW

The fat from cattle and sheep is known as tallow. Sheep fat is rarely used for edible purposes because of the difficulty of removing its strong flavour and odour. It is widely used for soap and candle making and in lubricants. Beef fat is used for all these purposes and for food as well. The fat from the heart, caul, and around the kidney, which corresponds to leaf lard, gives the finest edible tallow. Its best grades are used to produce oleo stock (also known in Europe as *premier jus*) for the oleomargarin industry by the same process which is used to produce neutral lard. Beef tallow of good quality is also used in so-called lard compounds to give a stiffer consistency. Tallow, except for oleo stock, is usually steam rendered. Inedible tallow is graded and sold on the basis of colour—sometimes also on the basis of its content of moisture, impurities, unsaponifiable matter, and free fatty acid. If sold for candle making, the titre test is especially important, since it indicates the yield of solid fatty acid suitable for candle making that may be expected.

GARBAGE GREASE PRODUCTS

The trade, therefore, often distinguishes between*packing-house grease* and all other greases. In small establishments doing a local business, the term grease may be applied to a product of variable origin and containing more or less tallow and fat recovered from retail butchers' scrap and the like.

Garbage grease is one of the more important of the waste fats. In some cities household, restaurant, and hotel offal is separated into garbage proper, consisting mainly of food offal, and into rubbish and trash of all kinds. Grease from the garbage is recovered usually by one form or another of steam rendering. It is sold mostly to candle makers. The disposal of garbage presents a serious problem to municipal administrators, and methods of rendering other than by steam, including solvent extraction, have been proposed.

Opinions differ concerning the value of different methods. The recovery of garbage grease is sometimes profitable, sometimes not, depending upon local conditions and the general level of prices of fats prevailing at the time the grease is sold. In some cities dead animals of various kinds are rendered with the garbage.

In others this is done by privately owned rendering companies which work up such material, and at times also the garbage from hotels and restaurants and the scraps from retail butchers' shops. The material produced by them is sometimes known as horse oil which may, however, and usually does, contain fat from other sources than the carcasses of horses.

PRODUCTION OF VEGETABLE FATS

Vegetable fats, except in special cases, are not produced by rendering. The principal exception is the palm oil produced by African natives in the home of the palm. They boil the crushed and more or less decomposed fruits with water and then skim off the oil.

The commoner method of producing vegetable fats is by expression. This process involves several steps. The first is the preparation of the seeds, which consists of the removal of the shell or hull. This decortication is not always necessary, as in the case of fruits like the olive or small seeds like rape or flax. Sometimes it suffices merely to crack or crush the shell without removing it. Decortication is usually done by special machines; in the case of the coconut, however, it is done by hand. The coconut presents a special case in another respect.

The material is now ready for the expression of the oil in presses of various types. This is sometimes done without heating the material, as in the case of olive oil. Such oils are known as cold-pressed oils. Sometimes the residue from the pressing is reground with or without the addition of water, again pressed, and a second portion of oil obtained. The first pressings of olives is known as virgin oil. If the olives from which it is made are of good quality, this oil represents the highest quality.

Since cold pressing does not extract all the oil, it is practiced only in the production of a few special edible oils, the natural flavour of which is highly prized. They are used without further refining other than clarifying and filtering. The oil unextracted by cold pressing is recovered in large measure by grinding up the residue from which the cold-pressed oil has been extracted, cooking it (usually with steam), returning it to the presses, and again expressing the oil. This is known as hot-pressed oil. The great bulk of vegetable oils—coconut, palm kernel, cottonseed, peanut, etc.—are extracted only by hot pressing. All these oils require refining before they are suitable for edible uses.

The expression of the oil is practiced either with presses or with special machines known as expellers. For descriptions of the construction of these machines, the reader is referred to special treatises. They operate discontinuously—that is to say, they act only upon one charge at a time. After the oil has been expressed from this the press must be unloaded and refilled with a fresh charge of the ground oil-bearing material. Hence the labour costs of operation may be considerable.

The expeller, on the other hand, operates continuously and the labour costs are correspondingly low, though maintenance and power charges are said to be high. The expeller is built on the same principle as the ordinary meat chopper or sausage machine which nowadays is to be found in most kitchens. The oil-bearing material is fed into one end of a cylinder within which a power-driven worm conveyor forces the material to the other end of the cylinder and out against resistance, exactly as though it were sausage meat. The pressure exerted in the process squeezes out the oil. In some factories the material, if very rich in oil, is first passed through an expeller to remove a part of the oil, then reground and recooked, and finally expressed in a hydraulic press.

The hydraulic press when skilfully operated removes the oil somewhat more completely than the expeller. However, neither does so completely. From 4 to 8 or even 10 per cent of oil may remain in the residue which is known as oil cake, sometimes also as press cake or pressed cake. A method has been devised to recover this oil. It is known as the solvent-extraction process. It consists of grinding the cake to a meal and then extracting or leaching it with a volatile liquid in which the oil is freely soluble. The extract is then drawn off from the cake into a still where the solvent is distilled off and recovered, leaving the oil behind in the still. The solvent most commonly used is benzene, although carbon bisulfid, petroleum products, and other liquids are also used. The oil obtained is inferior in many cases to that obtained by expression. It is especially so when the solvent used is carbon bisulfid; such oils are known as sulfur oils.

In Europe the solvent extraction of oil cake for the recovery of residual oil has been practiced quite extensively. In the United States it has been employed principally at times of high prices or for special purposes. The chief difficulty at present is that the ideal solvent remains to be found. Those commonly in use or proposed either introduce a serious fire hazard, or else in time they corrode the equipment, or finally they leave a bad odour or taste in the cake, which impairs its value as cattle feed. It has been proposed to do away with presses and expellers altogether and to dissolve out the oil from original oilbearing raw material by solvent extraction, and there are some plants of this kind in operation in Europe.

REFINING

Refining has for its object the removal of free fatty acids and other objectionable substances—principally nitrogenous and mucilaginous matters. The principle involved has already been outlined The melted fat or oil is treated with a little more than the requisite amount of strong aqueous caustic soda solution to convert the free fatty acid present into soap. The oil and the alkali solution are thoroughly stirred together and sometimes warmed. The mixture is then allowed to separate.

The result is that the oil freed from fatty acid floats on top of a layer of soap, alkali solution, and other impurities, which is drawn off. The oil is then washed with water to remove the soap, alkali, and other impurities, when it is ready for the decolorizing or deodorizing process. There are other methods of refining, but this is the one most commonly used in America.

The under layer of soap and other impurities, which is drawn off from the oil, consists of solid matter mixed with some water. It is known as foots, probably because it collects at the foot of the tank. A large proportion of it is soap. It may be sold to soap makers for use in the lower grades of soap, the price being based on the percentage of fatty acid present in it. Hence it is also known as soap stock.

Or it may be treated with strong sulfuric acid to set free the fatty acids contained in it. These then float to the surface, are skimmed off, and sold to the soap or candle maker. This product is known commercially as acidulated foots. It pays to produce it rather than to sell ordinary foots whenever the freight from the refinery to the soap maker is considerable, for by converting foots into acidulated foots the weight is reduced about one-half. A third use for foots is to convert it into washing powder by mixing with a suitable amount of soda ash. The soda ash takes up the water in the foots and crystallizes with water of crystallization, thereby converting the foots into a hard, dry cake which needs only to be broken up and ground to be salable as washing powder. At the same time the soda ash bleaches the foots and improves the colour.

While with the exception of virgin oils, Production of vegetable fats the great bulk of vegetable oils destined for food use is refined, this is not the case with animal fats. It was formerly the custom to refine the poorer grades of lard and of tallow to make them more suitable for edible purposes. The practice was to wash them in the molten state with a weak alkali solution, or to treat them with alum or other chemicals or with fuller's earth. Such refining is no longer permitted in federally inspected packing houses. About the only practice now allowed is to let the fat *settle* with the addition of some salt in order to remove traces of water and any fragments of tissue and fibre that may be present, or to treat with fuller's earth or other inert decolorizing agent. It is reported that in many regions of Europe refining of animal fats for food use is not prohibited.

FATS AND OILS TECHNOLOGY

DECOLORIZING OR BLEACHING

Most crude vegetable oils are deeply coloured. They must be bleached. There are many methods, but the one in most general use in America is to agitate them, after they have been refined, with some solid material which absorbs the colour. The usual material is fuller's earth. Since the war various forms of carbon and charcoals have also come into use, usually in combination with fuller's earth. This is the result of the stimulus to the production during the war of good absorbent carbons for use in gas-masks. To the dry oil dry fuller's earth is added, usually only a few per cent, the quantity depending upon the character of the oil and the temperature; the mixture is then warmed, usually to less than 80 degrees Centigrade, and agitated for one-half to one hour. It is then pumped through a filter press which retains the fuller's earth and permits the oil to run out clear. The canvas is usually backed by a corrugated metal plate so that it will not burst under pressure. In the filter press the frames are pressed so tightly together by means of a powerful screw that liquid cannot escape between them. The frames of the screens are perforated with a set of holes so placed as to form a continuous tube reaching from one end of the stack to the other when the screens are assembled. This tube opens into every other compartment.

Through this tube the liquid is pumped into these compartments and forced through their canvas walls into the adjacent compartments which do not communicate with the tube. The canvas holds back the fuller's earth and only permits clear oil to ooze through into the adjacent compartments. The latter are provided with channels which permit the clear oil to escape to the outside where it runs into storage tanks.) An appreciable amount of the oil remains in the earth retained in the press at the end of the operation. A considerable proportion of this is sometimes recovered by forcing dry steam through the press which carries out with it much of this residual oil. Nevertheless appreciable amounts of oil remain behind. It has been proposed to recover these by the solvent extraction process but this does not seem to be done at the present time in America. Since the necessity of using fuller's earth involves not merely expense but also loss of oil and since, therefore, the costs of decolorizing rise with the amount of earth it is necessary to use, the price paid for an oil, and especially for cottonseed oil in the United States, depends among other factors upon the ease with which it may be decolorized or bleached. Furthermore, if too much fuller's earth has to be used, the oil acquires an earthy flavour.

DEODORIZING

Many oils even after they have been refined and decolorized to transparent whiteness retain a disagreeable odour and flavour. This is especially true of cottonseed oil. Such oils may be deodorized by blowing steam through them, since the substances responsible for odour and flavors are usually volatile. A still more effective way is to blow the steam through the oil after it has been heated to a high temperature, say 340 deg Fahrenheit. The most effective method is to carry on this treatment in a vacuum.

WINTERIZING

Now, different triglycerids solidify and melt at different temperatures. Therefore if an oil consists of a number of triglycerids, some of which remain liquid at low temperatures while others become solid, and such an oil is exposed to a low temperature, more or less sediment forms consisting of the triglycerids that separate out at that temperature. Indeed, the oil may be completely converted into a solid cake if it is cooled down far enough.

Every housewife who has permitted a bottle of olive oil to stand outdoors in winter has made this observation. This solidification was especially objectionable when fatty oils were used in lamps. It is objectionable in salad oils today because the housewife, being ignorant of the nature of the phenomenon, is apt to believe the oil spoiled. Manufacturers, therefore, subject such oils to a process which prevents the separation of solids in all but the most extremely cold weather. An oil so treated is known, naturally, as a *winter* oil, and the process is known as *winterizing*.

Conversely, an oil that has not been *winterized* may be known as a *summer* oil. Winterizing is a very simple procedure. The oil is very slowly chilled in large tanks to the temperature at which it is to remain clear. It is allowed to stand quietly at that temperature for a considerable length of time to permit the separation of the solid crystalline materials from the liquid to become complete. The oil with the suspended solid matter is then pumped through filter presses which retain the solid and allow the liquid to run out. The solid remaining in the press is known as stearin because largely composed of the glycerids of stearic acid. It is used to stiffen lard compounds, and in soap and candle making.

PRODUCTION OF STEARIN

The amount of stearin obtained by winterizing oils is small and would not supply the demand for it. Therefore, not merely oils but also solid fats are treated to separate them into a more solid and a more liquid fraction. The fat is melted and allowed to cool slowly in large tanks to a given temperature which is so chosen that the solid portion which gradually separates has the desired melting point.

The fat is held at this temperature for some time in order to permit the solid portion to form completely. Cooling slowly and prolonged holding have another effect. The solid portion is caused to separate in coarse granular masses rather than in fine particles which could not so easily be freed from the liquid portion. Hence this process is known as *graining.*

When it is judged that the mixture has the desired granular texture, it is placed in a powerful press and the liquid portion squeezed out and thus separated from the solid portion, the stearin. In this manner are prepared lard stearin and lard oil (usually from grease, less nowadays from lard), and oleostearin and oleo oil from edible tallow. Some tallow stearin and tallow oil is also produced from inedible tallow. Lard stearin, if produced from lard, is used especially to mix with other lard that is destined for a warm climate to stiffen it. Such lard is known in America as *Cuba lard.*

Lard stearin from grease is used for soaps, candles, and lubricants. Lard oil was formerly widely used as an illuminant instead of sperm oil. Though it is still so used to a very limited extent, it has been displaced nearly entirely by petroleum products. Lard oil is used in compounding lubricants and especially as a so-called cutting oil, *i.e.,* an oil used to lubricate the cutting edges of steel tools in metal working. Oleostearin is used principally in lard compounds, in shortening agents, while oleo oil is used in oleomargarin and for shortening.

HYDROGENATION

As it happens, practically all the important unsaturated triglycerids, *e.g.*, triolein, trilinolin, and trilinolenin, are liquids at ordinary temperatures; hence fats which contain them in considerable proportions are oils or, if the proportion is smaller, they are soft solids. On the other hand, the important saturated triglycerids, *e.g.*, tristearin and tripalmitin, are solid, and fats which contain them in preponderating proportion are firm at ordinary temperatures. Now, since hydrogenation converts unsaturated triglycerids into saturated ones, it changes oils into solid fats.

This process is of the greatest commercial and economic importance, since it permits of the conversion of oils into fats and thereby widely extends the substitutibility of oils for solid fats and even of oils for one another. In practice, hydrogenation is widespread, but it is used far more for edible than for industrial products, since the cost is of the order of magnitude of one-fourth to one-half a cent a pound, which may be prohibitive for many industrial uses. Moreover, in the inedible field the possibilities of substitution of one natural fat or oil for another are greater than among edible products. Nevertheless, the introduction of hydrogenation has had a profound effect upon the fat and oil industries and trade of the world, for not merely has this discovery widened the uses of oils but the hydrogenated product is usually improved in keeping quality and in colour, odour, and flavour as well.

The commercial process of hydrogenation is based upon the purely scientific researches of the French chemist Sabatier and of his students. Though they discovered the scientific principles, commercial application of them was made by others who have taken out a host of patents. The principle itself is simple. To the perfectly dry oil is added a small amount of very finely divided nickel or compound of nickel, called a catalyst.

(A catalyst or catalyzer or catalytic agent—the terms are synonymous—is a substance which affects the velocity of a chemical reaction without itself appearing in the final product. In the present case nickel is the catalyst. It speeds up the rate at which the oil absorbs hydrogen and may be recovered in undiminished amount at the end of the reaction. A catalyzer cannot start a reaction; it merely modifies the velocity of the reaction.

A large quantity of the reacting substances can be transformed by a very small quantity of the catalyzing agent.) Other metals may be used, but nickel is the one most widely employed. The oil with the nickel suspended in it is placed in a tight, strong metal vessel and heated. At the same time pure hydrogen gas is forced into it until a definite pressure is reached.

The vessel contains, commonly, some mechanical device to churn up the oil as the gas passes in so that all parts of the oil may be mixed intimately with the hydrogen. The process is interrupted when a sample of oil withdrawn from the vessel is found to have the desired properties. The oil is then withdrawn and cooled sufficiently to permit of its being filtered to remove the suspended nickel.

SUBSTITUTABILITY AS A TECHNOLOGICAL OBJECTIVE

Hydrogenation, refining, deodorizing, decolorizing, stearin pressing, winterizing, and a large number of other technological processes which have been evolved in the course of the last 150 years have all had one object: to make one fat substitutible for another. These substitutions have at different periods had different purposes.

When candles were important and hard fats were in demand, the preparation of a hard fat from a soft one by separating the stearin was discovered. When with the development of lamps burning oils became more important, the same method of pressing stearin made fluid oils derived from harder fats available. When with the development of the production of vegetable oils these became more abundant than solid fats, lard compounds were developed, facilitating the use of oils as cooking fats in countries where hard cooking fats are preferred to oils.

This was followed by the introduction of hydrogenation which widened the substitutibility of oils for hard fats. It still remains for chemists to discover a commercial method of converting a saturated fat into an unsaturated drying oil; but even without it the most striking characteristics of the evolution of fat and oil technology have been to increase greatly the possibilities of substitution of one fat for another. If substitution is not more widely practiced, the deterring factors are price, cost of the treatment, and finally the fact that despite the great progress in the treatment of fat it is as yet not possible to modify all fats so as to give each and every one of them the peculiar properties of every other.

SOAP MAKING

There are many methods, but they all may be divided into two groups. Soap is formed either in one operation or in two. If it is done in one operation, the fat is simply treated, usually hot but sometimes cold, with the appropriate amount of a solution of caustic soda or potash.

This forms soap and glycerin. There are a number of methods of separating the soap from the water, glycerin, the excess of alkali, and impurities, but the commonest is simply to add a considerable amount of ordinary salt. This dissolves in the water present and forms brine. Now, soap is but slightly soluble in strong brine. Therefore, the mixture separates into three layers: an upper layer consisting of the purer portion of the soap; a middle layer, dark in colour, consisting of impure soap and known as nigre (from the French *nègre*, black); and a bottom layer of brine containing glycerin. The upper layer is run into molds or otherwise formed into the well-known commercial soap units. The nigre is worked over and purified in various ways and finally worked into soap. The brine may be run into the sewer ultimately or the glycerin may first be recovered from it.

CANDLE MAKING

Up to the early part of the nineteenth century candles were made from beeswax, spermaceti, or tallow. Those made from the first two were the firmest and gave the best light; but they were also expensive and became more and more so with the decline of whaling, for spermaceti is a solid wax which separates from the oil obtained from cavities in the head of the sperm whale.

To meet the scarcity of beeswax and spermaceti, candles began to be made from stearin. By the 1830's, however, they began to be made from the solid fatty acids obtained in the saponification of fat, and this process has remained important to the present day. It consists in splitting tallow, grease, stearin, or any mixture of them into glycerin and free fatty acids by any suitable method. Thus a mixture of free fatty acids is obtained which, like the fats from which they were produced, consists of both more fluid and more solid acids. The solid portion is separated from the fluid portion by cooling, graining, and pressing, much as lard stearin is separated from lard. This solid portion is known in the trade as *stearic acid* and the candles made from it as *stearic* or *stearin* candles, although it is by no means pure stearic acid. It is nearly always a mixture of palmitic and stearic acid, as well as of any other solid acids that happen to be contained in the raw materials from which it was produced. The oil which is obtained in expressing the *stearic acid* is known as *red oil.* It consists mainly of impure oleic acid and is used either in soap making or as soap for use in woollen and other textile mills. From the *stearic acid* candles are made by melting and casting in suitable molds. To overcome their brittleness they are usually mixed with some other material such as paraffin. Indeed, paraffin has to some extent displaced *stearic acid* in candles.

The candle maker in purchasing his raw materials is primarily interested in the amount of solid acids they contain, for it is these that he wishes to use in his candles. For this reason he prefers fats with a high content of solid fatty acids. As the titre test (II. Properties of Fats and Oils—Other useful tests) is an index of this, it is of especial value to him. Since the content of solid fatty acids is usually comparatively high in tallows and stearins, he prefers these fats to greases and oils. In buying raw materials he is also interested in their colour, for from dark greases are obtained dark acids from which white candles cannot be made directly. The acids must either be bleached or distilled. The latter is the preferred practice. The common fatty acids are not readily volatile at the pressure of the atmosphere. When distilled under ordinary conditions they char and a great proportion is converted into a sort of pitch. If, however, the distillation be done in a high vacuum, there is but little decomposition and but little pitch is formed. The acids that distill over are pure and white even though the grease from which they are produced may have been very dark.

LUBRICANTS

Until the rise of the petroleum industry the most important lubricants were tallows, greases, and vegetable oils. For about seventy-five years now petroleum products have gradually been displacing them, but this displacement has by no means been complete. Material quantities of fats and oils are still used in lubrication, usually mixed with greater or lesser proportions of petroleum derivatives. It is therefore worth while to consider the manner in which fats and oils are used in lubrication. Only non-drying oils are suitable, for drying and semi-drying oils absorb oxygen and become sticky and gummy. Fats are still used in so-called cylinder oils, used to lubricate the pistons within the cylinders of steam engines.

They are no longer used alone but mixed in relatively large and increasing proportions with mineral oils. Tallows, greases, and lard oil are used in all types of engines except marine engines, in which compounds containing rapeseed oil are the only ones acceptable. However, not only is the percentage used on the decline but the total amount of cylinder oils consumed is diminishing relatively because of the displacement of the steam engine by the steam turbine, the internal combustion engine, and the electric motor.

Lubricating greases, also extensively used for automobiles as well as for other machinery, consist of mixtures of mineral oil, animal grease or tallow, vegetable oil, and soap made from such grease, tallow, or oil. The soap apparently serves not merely as lubricant but also to emulsify the mixture so as to give it the desired consistency. Such greases represent one of the more important lubrication uses of fats.

Another considerable use of fats and oils for lubrication is in metal working, to lubricate the cutting edge of tools. Lard oil is perhaps the preferred oil for this purpose. In recent years, with the development of high-speed tools, so-called soluble oils are being used. These consist of mixtures of animal or vegetable oil, mineral oil, and alcohol with some other minor ingredients. Soluble oils are not used directly but are first mixed with ten to twenty times their volume of water.

MARGARIN

Margarin is designed to furnish a substitute for butter. It was invented in 1869 and was originally a purely animal product. It is usually made today either from a mixture of animal and vegetable fat or from vegetable products alone. The most important materials are neutral lard, oleo stock (*premier jus*), oleo oil, coconut oil, palm kernel oil, peanut oil, cottonseed oil, oleostearin, lard stearin, and sesame oil. Any two or more of these are mixed together so that the mix has a melting point from 26 to 27.5 deg Centigrade. The different formulas used vary greatly and sometimes depend upon the price relations of the different fats.

The mixture chosen is melted. It is then cooled and mixed with skim milk *ripened* or soured with a pure culture of bacteria. The mixture is agitated to emulsify the fat and then cooled by causing it to fall against a spray of ice-cold water which carries it into a tank of ice water from the surface of which it is skimmed off. It is thereafter handled exactly like butter. In Europe, in some factories, it is cooled on a cooling drum instead of with ice water, but apparently this machine is not used in the United States. Margarins, then, are mixtures of fat emulsified in skim milk and made into the semblance of butter. In the United States there are two main types: one consists usually of neutral lard or oleo oil and cottonseed oil; the other of coconut oil and peanut or cottonseed oil. Hydrogenated fats may also be used, usually so-called hydrogenated coconut oil, which is usually a hydrogenated mixture of coconut oil with a small proportion of peanut or cottonseed oil.

LARD COMPOUNDS

The production of lard compounds began in the late 1870's as adulteration of lard with tallow or beef stearin. Soon thereafter, cottonseed oil was also used as an adulterant. The proportions of tallow or stearin and of cottonseed oil were gradually increased until they formed the major constituents and lard the minor ingredient.

By 1890 brands were on the market which contained only enough lard to give the characteristic flavour. Up to that time these products were sold as refined lard, pure family lard, etc. About this time, owing to a Congressional investigation, manufacturers began to brand products of this kind as lard compounds, the designation by which they have been known ever since.

About 1908, shortly after hydrogenation began to be practiced in the United States, a new type of cooking fat began to be introduced which consisted of cottonseed oil hydrogenated to the desired consistency. Products of this kind have not usually been marketed as lard compounds but under their own distinctive brand names.

Today there are three principal types of lard compound on the market:

1. The original type of lard compound consisting of beef tallow or beef stearin and a vegetable oil, preponderatingly cottonseed oil. It may also contain some lard or lard stearin.
2. A mixture of cottonseed oil more or less completely hardened by hydrogenation and a vegetable oil, preponderatingly cottonseed oil.

3. Cottonseed oil partially hydrogenated to the desired consistency.

The method of manufacture is relatively simple: The component fats, so proportioned that the mixture has the desired melting point, are melted together. The resulting liquid is run onto hollow revolving cylinders chilled from within, known as lard rolls. The mixture is thus rapidly chilled, thereby acquiring the texture and appearance of lard. The chilled fat is automatically scraped off and drops into a trough in which a worm conveyor beats it up and transports it to storage tanks from which it is packed into the shipping containers.

CONDITIONS AND TRENDS OF PRODUCTION

INFLUENCE OF AGRICULTURAL EVOLUTION

Before that time sheep and cattle raising was the only form of agriculture possible in vast regions of grass lands because of scarcity in these lands of fuel, building materials, means of transportation, and water. In such regions carcass fats, together with hides, were the major marketable products of agriculture, for these are not very perishable. Meat, because superabundant and perishable, had little value. Cattle were slaughtered for their hides and tallow; sheep were sheared, or were killed for their pelts and tallow. It was not so long ago that this was the practice in the Southwest of the United States, in Argentina, and in Australia.

It is still the marketing method in some remote regions, for example, in certain sections of South America. As late as the 1850's, probably even in the 1860's in Chicago, Cincinnati, and elsewhere in the Middle West, only the hams and shoulders of a large proportion of the hogs slaughtered were marketed as meat. The remainder of the carcass was steam rendered for lard. As late as the 1870's cattle were slaughtered for their hides and tallow in California.

In other words, in certain stages of agriculture and in certain remote regions hides and carcass fat are the major products of animal husbandry; meat is either worthless or a mere by-product. The world over, the area of grass lands that was producing carcass fat as a major product was so vast during perhaps the first three-fourths of the nineteenth century as to exercise a great influence on animal husbandry everywhere, as well as on the world trade in fats and oils. The hides and wool were the first products, the fat the second product, the meat the last product.

Gradually, however, the grass lands were brought more and more under the plow. Three sets of inventions made the growing of crops possible in these areas. One was the railroad, which not merely made the transportation of crops to market practicable but also served to bring to the farmer fuel and building materials in exchange. Later came the modern steamship, port storage facilities, and still later the present-day refrigeration. On such land a farmer must work a considerable acreage to produce enough to support a family. If he must work the land with a spade or a primitive plow and a bullock he cannot cultivate a sufficient area. To exist he must practice extensive agriculture upon a considerable acreage, but this requires machinery operated by an abundance of draft animals or by engines of one sort or another.

ANIMAL FATS AS BY-PRODUCTS

There are now left few regions of the earth where hides and carcass fats remain major products of animal husbandry. In most regions, as a result of a gradual shift in relationships, meat has become the major product and fat a by-product in the sense that much the greater return is received from the meat and the lesser return from the fat and other products. The capacity for meat production has not yet reached its limits.

The other aspect of the question is more obscure, yet perhaps more important. Meat has probably always been the primary product of animal industry in regions close to centers of population. As improved transportation and refrigeration have brought to these centers of population the meat as well as the fats produced in remoter regions, it might be supposed that in the centers of population prices of meats should have fallen relative to prices of fats. But on the contrary it is the prices of fats that have declined. The reasons are many; the decline in prices of fats relative to prices of meats in the centers of population is the resultant of a number of factors, some involving supply, some demand. On the supply side, one factor has been the development of improved breeds of animals, particularly cattle, capable of laying on more fat. These heavier fat-producing breeds have been adopted chiefly because of the better quality of meat produced, but a consequence has been a larger production of fat per pound of meat.

The supply of animal fats is strongly influenced by the demand for meat, since it is the price of meat which is chiefly influential in determining the profits of animal husbandry. The direction and the amount of these changes over the last 50 years are by no means clear. It appears that in the United States the demand for meat per capita probably declined between 1907 and the beginning of the war and has since increased slightly. Even for the United States and for these recent years, the data do not justify positive conclusions, and for other regions and for other years there is too little evidence to justify even a guess as to the direction in which the supply of animal fats has been influenced by the demand for meats.

On the demand side, there has been an important increase for fats in general, arising from the greatly increased consumption of soap. The most powerful influence on the supply side, however, has been the development of new sources of vegetable fats. In consequence, vegetable fats have been substituted for animal fats on a large scale and in many uses. The production of cottonseed, coconut, and palm kernel oil has been greatly expanded.

As the demand for cotton has increased, production of cottonseed has of necessity increased likewise. New areas opened up in the tropics have provided abundant supplies of coconut and palm kernel oil at low prices. With the increase in supplies of vegetable oils came also the development of the process of hydrogenation, whereby the vegetable oils could be transformed into solid fats and brought into direct competition with the animal fats. The relation of meat (in the stricter sense) to fat can be shown both by comparisons between animals of different ages and by comparisons of the prices of different parts of the carcass. Let one compare the bacon type of hog with the lard type of hog.

The bacon type of hog is finished at an earlier age, commands a premium in the livestock market, and when the disposition of the entire carcass is computed, it is observed that the lard and salt pork fractions are relatively low. In the case of the lard hog, the animal is finished later and reaches a larger weight; the heavier the animal, the lower the price per pound as a rule; and when the disposition of the carcass is computed, it will be found that the lard and salt pork fractions are relatively high.

Table. Prices of Hogs and Hog Products in Chicago

Year	Live Hogs	Smoked Hams	Bacon: Short Clear Sides	Fresh Pork Loins	Pure Lard	Short Ribs	Mess Pork
1913	8.35	16.6 (199)	12.7 (152)	14.9 (178)	10.8 (129)	11.37 (136)	10.81 (129)
1914	8.30	16.7 (201)	13.2 (159)	15.4 (186)	10.2 (123)	11.09 (134)	10.44 (126)
1921	8.51	26.8 (315)	13.5 (159)	22.5 (264)	13.2 9.52	(155) (112)	10.75 (126)
1922	9.22	26.5 (287)	14.1 (153)	21.7 (235)	13.1 (142)	11.21 (122)	12.37 (134)
1923	7.55	21.2 (281)	12.0 (159)	18.0 (238)	13.9 (184)	9.81 (130)	11.95 (158)
1924	8.11	20.2 (249)	14.4 (178)	19.1 (236)	14.7 (181)	11.28 (139)	13.14 (162)
1925	11.81	27.1 (229)	22.3 (189)	25.0 (212)	17.9 (152)	16.97 (144)	18.79 (159)
1926	12.34	30.8 (250)	20.1 (163)	27.8 (225)	16.9 (137)	15.48 (125)	18.30 (148)

The disparity in prices is so pronounced as almost to justify calling the lard a by-product. The high price of the lean cuts must carry the low price of the salt pork and lard; and since the load is heavier in the case of lard hogs than of bacon hogs, this represents an advantage for the latter, considered from this viewpoint alone.

This conclusion is confirmed by inspection of the practices current in the retail sale of fresh pork. The loins, chops, and steaks—the cuts that are prepared by broiling rather than by roasting—carry the highest prices. Next come the relatively lean cuts that may be either broiled or roasted, such as the rump. Then come the fatter cuts that can be neither broiled nor roasted, but must be boiled, stewed, or sautéd. As meats are sold at retail, the excessive fat is cut off and thrown away, to be rendered or sold to renderers for what it will fetch. In the household a further trimming off of fat occurs, representing largely a waste. Owing partly to preference for the quality of meat in the parts of the animal sold as steaks and roasts, but largely also to the co-existence of the fat, the lean cuts carry a higher price and the fat cuts carry a lower price.

In the case of beeves, it is a fair statement of retail practice to say that the hind quarters must carry the front quarters, and that in the case of each of these a relatively small fraction of the leaner, more highly esteemed cuts (steaks and roasts), carries the balance of the quarter. Lamb is preferred to mutton because it is leaner and younger. The premium for young, lean poultry over old, fat poultry expresses the same fact. Right through animal husbandry (outside of dairying), therefore, we find the fat to be at best a subordinate product, if not indeed a by-product; the lean is the principal product.

VEGETABLE OILS MAJOR PRODUCTS

If now the vegetable fats and oils be compared with carcass fats in regard to manner of production, it appears at once that the former are not in many cases joint products. They are usually major products, sometimes with and sometimes without by-products of value. It is instructive to review the different forms.

In the case of palm oil, this is practically the sole product, since the carbohydrate and protein in the residues left after the extraction of the oil find very little use in the country of origin. (*Carbohydrate* is the term applied by chemists to a group of substances which includes the sugars and substances which when decomposed by hydrolysis.

The most important representatives are grape sugar or glucose, fruit sugar, milk sugar, malt sugar, cane or beet sugar, starch, cellulose—which last is the most important constituent of wood, cotton, and many other fibrous plant products. *Proteins,* also known as albumins or albuminous substances, are a class of complex chemical substances present in all living things. They are familiar to everyone for, with water, they are the chief ingredient of the white of eggs, of flesh (muscle tissue), and of blood clots.

They always contain carbon, hydrogen, oxygen, nitrogen, usually sulphur, also sometimes phosphorus and iron. Very few of them are soluble in anything but aqueous media. Most of them are coagulated by heating (for example the albumen in white of egg), that is, heat makes them insoluble.) In the case of the palm kernel and coconut, these may be crushed in the countries of origin or in countries of destination. When crushed in countries of origin, the residual cake finds little use, except as it may be shipped to other countries for feeding purposes. When crushed in the countries of destination, the cake and meal are used as high-protein feeding stuffs and sell in competition with other high-protein feeding stuffs.

In some importing countries, the protein of the oilseeds is so highly prized as a feeding stuff that the oil cake has very considerable value. But if crops be viewed as a whole, the oil of the palm and the fat of the coconut are the principal products, and the carbohydrate and protein of the residues are the by-products. In the case of the soy bean in Asia, on account of the importance of the whole bean in the diet of the common people, it is reasonable to term the fat, the protein, and the carbohydrate co-equal products; when soy beans are crushed in countries of import, however, the fat is the principal product and the cake is the by-product. The same thing holds for the seeds of the peanut, cotton, sunflower, and for the olive—the fats are the prime product and the residues the by-product, although cottonseed oil is merely a by-product in the production of cotton and corn oil a by-product in the manufacture of meal and starch.

Quite generally with oil seeds, therefore (cottonseed oil and corn oil being the exceptions), the fats are the chief products and the carbohydrate and protein the by-products, varying under different circumstances from a relatively high value to insignificance. Furthermore, in the case of animals, the meat (chief product) is made to carry the load, whereas in the case of oil seeds, the fats and oils (chief products) must carry the load of the commercial operation.

RELATIVE COSTS

Comparison of costs of production of animal and vegetable fats involves numerous complicated problems and cannot profitably be carried beyond a certain point. Certain of the facts bearing on costs are, however, pertinent to the discussion of conditions and trends of production. Fats arise in animals either through the storage of fat ingested with the feed or through the transformation of other feed constituents, of which starch is the most important.

These processes are accomplished with losses both of material and of energy, the absolute magnitude of which varies with circumstances. As a matter of fact the losses are very considerable and only a portion, sometimes only a small portion, of the intrinsic caloric values of the feed is recovered as fat and meat in the carcass. From the agricultural or commercial point of view, the losses incurred when feeding stuffs are converted into carcass fat may or may not represent waste and inefficiency.

This depends upon the value of the feeding stuffs and the conditions under which these were produced. If the feed is fit for human food and might have been used for this purpose (as, for example, barley, oats, corn, wheat), or if it was produced upon land that might have been devoted to producing food crops, then the production of carcass fat in this manner might be regarded from one point of view as inefficient, even though it may have been commercially profitable, because the carcass fat and the meat will feed fewer persons than might have been maintained upon the grains used as feed to produce them. It will be advantageous to illustrate the limited recovery of the calories of nutrients in the flesh of animals raised on them and to indicate the relative preponderance of fat over protein with increasing age of the animals. With good practice in the corn belt in this country, 8 or 9 bushels of corn are required to produce a hundredweight of live hogs, corresponding to a recovery of about a third of the calories of the corn in the form of edible pork.

In the feeding of adult steers, the recovery of the calories of corn in the form of beef is often as low as 15 per cent. Possibly 45 per cent in the case of hogs and 30 per cent in the case of cattle represent the maximum limits of practicable recovery of nutrients in the form of edible meat products. To illustrate the preponderance of fat over lean with increasing ages and weights, information may be drawn from analyses of whole carcasses. Hogs weighing in the neighbourhood of 200 pounds will contain something like 24 pounds of protein and 85 pounds of fat, while hogs weighing in the neighbourhood of 300 pounds will contain some 30 pounds of protein and 160 pounds of fat. From these data obvious inferences as to what might be termed physiological costs of production may be deduced; but whether physiological costs of production correspond to economic costs of production depends on other circumstances.

Animal fats, then, are not primary products of the soil, they are secondary products. They are produced by feeding primary products of agriculture. The conversion of these primary products, the feeding stuffs, into animal fats is accompanied by heavy losses in nutrients. It is quite otherwise with vegetable fats; these are primary products of the soil, and no such conversion losses are involved. If animal fats were produced only from vegetable fats of the feed one might be justified in concluding offhand that animal fats are always produced at higher cost than vegetable fats. As a matter of fact they are produced in this manner to a slight extent only. They are produced for the most part by conversion of non-fat material in fodders like alfalfa, hay, straw, stover, bran, and especially from the starch of feed grains. There are still regions where bulky products like fodder and grain have little value because of transportation difficulties, where oilseed crops cannot be made to flourish. In countries like the United States, however, except for range cattle and sheep, animal fat is largely produced from the grains, particularly corn.

Despite the losses involved in converting plant nutrients into animal fats, it is more economical in temperate regions to obtain fats chiefly by this indirect process than directly from plants. This is the case for two reasons. Because meats cannot be produced without the simultaneous production of fat, a large amount of animal fat is unavoidably produced in the process of obtaining the supplies of meat for which we are willing to pay. A large proportion of this fat would be produced even though it had no more remunerative use than to supply heat for the power plants of the packing houses. Because the fat does have a considerable value, however, it is profitable to produce more than the theoretical minimum per pound of meat. The longer an animal is fed, the larger the proportion of nutrients converted into fat and the smaller the proportion converted into lean meat. With the decrease in. value of fat relative to meat in the last generation, it has become necessary to shorten the feeding period and thus reduce the ratio of fat to lean meat in the carcass, but this ratio is still far above the minimum.

No highly efficient fat-producing plants are adapted to cultivation in most temperate regions. In consequence, vegetable fats can be produced economically in temperate regions only in connection with other joint products capable of carrying the major burden of the costs of production. The necessary conditions are those which account for the existing production of cottonseed and corn oil and to some extent of peanut oil. With certain tropical plants, however, the situation is otherwise. The coconut and the palm store large quantities of nutrients in the form of fats. They are therefore fairly efficient direct fat producers. With this efficiency, coupled with the fact that they grow in regions where there are no other highly profitable uses for the land, the coconut and the palm are able to compete in fat production with the animals and plants of temperate regions where the major portion of the costs of production are borne by other joint products.

CONDITIONS AND TRENDS OF PRODUCTION

POSITION OF DAIRYING

Dairying is in quite a different position from other forms of animal husbandry, for several reasons. The cow is more efficient in converting feed into human food, in the form of milk, than is the steer or the sheep or even the hog. In addition to milk she produces veal as a by-product and is herself in the end turned into beef when she loses her efficiency as a producer of milk. While the hog stands closest to her in efficiency as a converter of feed into food, she has the great advantage over the hog that she thrives on fodder which is not suitable food for man, whereas much of the diet of hogs must consist largely of grains which are fit for human food. Finally, the dairy cow produces two food elements that are highly prized and therefore high priced.

One is protein of high quality; the other is milk fat which in milk, cream, and butter carries a premium price that places it in a class by itself as compared with other fats. Because of its preferred position, butterfat in the United States has not felt as yet to any material degree the competition of the domestic or imported vegetable oils. To be sure there is some competition from the margarins, but unless existing conditions and legislation change greatly it is not likely soon to become much more severe. The consumption of margarin, about two pounds per capita per annum, is small as compared with the consumption of butter, about 20 pounds per capita per annum. Moreover, an appreciable fraction of the margarin consumed is used as a superior cooking fat and here competes rather with animal and vegetable shortenings than with butter.

Dairy farming, then, is in a favourable position because it produces a fat that is so highly prized that it stands to a considerable extent above the competition of other animal and vegetable fats. In addition it produces protein of the highest quality. Where the dairy farmer is so located that he has a market for whole milk he gets some return for the excellent protein his cows produce. This is especially the case where there is a market for whole fluid milk as in the neighbourhood of towns and cities. Formerly this was the only way the dairyman got much of a return for the protein of milk. Elsewhere milk was paid for on the basis of its butterfat content, for the major use was in the production of butter, the skim milk being wasted or fed to hogs and poultry. Even when milk was used for the making of cheese, which contains most of the protein of milk, the price received was based on the price of butterfat because cheese and butter factories competed with one another.

This situation has begun to change. More and more milk is marketed as whole milk because the percentage of the country's population that is living in towns is increasing and because there is an upward trend in the per capita consumption of fluid whole milk. The development of the condensed and evaporated milk business has a similar influence. But a great effect has come from the development of the powdered milk and concentrated buttermilk business. Buttermilk contains much valuable protein, milk sugar, and mineral salts. Formerly, as already stated, it was fed to hogs and poultry so far as local conditions made possible, and the rest was wasted. Much of such buttermilk is now condensed.

The product is concentrated so that it can stand transportation charges; its keeping quality is such that it can be stored. In consequence it finds a wide market as feed, particularly for poultry. Powdered milk is mostly skim-milk powder made from skimmed milk, a by-product in the manufacture of butter. It contains all the valuable food elements of milk except the fat and is an excellent human food. It is widely used by bakers, confectioners, and chocolate manufacturers. It is saving for human food purposes many millions of pounds of milk protein and sugar that were formerly wasted. As these uses of buttermilk and skim milk grow, the net result ought to be a better return to the dairyman for his milk protein. He can therefore look forward to a steady strengthening of his economic position, and we may look forward not to a decreasing but to an increasing butter production, to be followed perhaps some day by a decline in butter production because of an insistent demand for whole milk.

METHODS OF PRODUCING VEGETABLE FATS

Reference has been made repeatedly to the competition between animal and vegetable fats. It has been pointed out that, irrespective of the relative costs of production of animal and vegetable fats on the same farms, it is more profitable for the farmer in countries like the United States to aim at the production of meat rather than animal fat because of the relatively high price of lean meat as compared with animal fat.

It was pointed out further that it is really immaterial in the industralized temperate zone what the cost of producing vegetable fat is. The competition which animal fats have to face in these regions is not with vegetable fats domestically produced as primary products, but with domestic by-product oils like cottonseed and corn oils and with oils of foreign origin. It is therefore of interest to examine in what manner these competing vegetable fats are produced.

Vegetable fats are the products in part of annual plants, in part of perennial trees. Among the annuals we have cotton, flax, peanuts, soy bean, sesame, rape, sunflower, and corn. Olive and tung trees are grown in orchards; coconut palms and oil palms in groves, some native, some planted. The planting of annuals is adjustable to demand. Trees require time to come into bearing, and have thereafter a varying period of bearing.

In the production of the fats of the palm and coconut there is, so to speak, much more of nature and much less of man than in the production of fats and oils in the temperate zone. Originally palm and coconut fats were secured from native growths and the elements of costs were largely those of labour and transportation. Latterly, plantation development of tropical oil seeds has come to pass, enlarging the producing area. The labour requirements are widely different from those in temperate zones and the labour is of different type. Though some groves are cultivated, for the most part palm nuts and coconuts receive only harvesting and preparation for the market.

The oil-bearing plants of the temperate regions (apart from olive and tung) must be planted, cultivated, and harvested, often when other crops compete for labour. Moreover the labour costs vary widely with the different vegetable oils produced in temperate regions. We lack anything like accurate or comparable cost data for both tropical oils and oils of the temperate zone.

Indeed for such by-product oils as cottonseed and corn, joint cost data necessarily would have to be arbitrary. In addition varying transportation costs, greatest for tropical oils, varying costs of extraction from the raw material, and varying costs of refining would need to be included in the calculation. A definite assertion concerning the relative production costs would therefore be unjustifiable. All that is justified, based on methods of production, is that it seems very probable that the outlays incident to production are materially lower for tropical fats than for vegetable fats produced in temperate zones. This finds expression in the decline of oil-seed crop culture in industralized countries except when it is a by-product like cottonseed or corn oil.

RESPONSIVENESS OF PRODUCTION TO PRICE CHANGES

Hitherto in the discussion of the competition between the various fats and oils only factors affecting long-time trends have been mentioned. There are, however, certain differences in methods of production that cause the several fats to react in different ways to abrupt changes in price levels. Production of some fats is more responsive to such stimuli than that of others. Hogs, for example, are more rapidly maturing animals than cattle. Moreover they multiply more rapidly. Any rise in the price of lard can be met more rapidly by increasing the swine population than a rise in the price of tallow can be met by increasing cattle population. Furthermore, for the same reason, the rapid growth of hogs, the lard supply can be increased more rapidly than the beef-tallow supply by feeding the hogs more heavily and marketing them fatter. Indeed, lard in the United States stands in a peculiar position, since together with hog meat it represents a major method of sending corn (maize) to market. The response to any stimulus to increase or diminish the lard supply must therefore be the resultant of at least three factors: the price of lard, the price of corn, and the number of hogs available.

The relationships are complex, and will be analysed in a subsequent study in this series. In the United States, moreover, carcass-fat production is responsive to changes in the general price level of fats to only a limited degree since, as already pointed out, the production of carcass fat under present conditions is not an independent enterprise but incidental to meat production.

The response of production to price stimuli must therefore be comparatively feeble. Mutton tallow stands in a somewhat different position from lard and beef fat, for lamb and wool together are the major product of sheep raising and tallow is the by-product. The price of tallow affects the production of mutton tallow far less than do the prices of lambs and wool, indeed scarcely at all.

Among the vegetable fats, cottonseed oil and corn oil are in much the same position as the animal fats, the price of the fat being only a minor factor in determining the production. The production of other vegetable fats, however, is quite sensitive to changes in price levels. Those fats that are obtained from annual crops may be made to increase or diminish in supply by expansion or contraction of acreage. Such fats are linseed, peanut, rapeseed, sesame, sunflower, and soy bean oils. Fats obtained from perennials, usually trees, are naturally not subject to rapid changes in acreage.

Trees take too long a period to come into bearing, and once arrived at this state they represent too heavy an investment to be displaced by another crop unless the producer has become convinced that fat production has become indefinitely unprofitable rather than temporarily so.

The principal fats from trees are coconut, palm kernel, olive, and tung oils. The supply of these oils varies for one of two causes. In the case of palm, palm kernel, coconut, and perhaps also tung oils, the nuts are simply not gathered if the price is not remunerative. Coconut and oil palms, being tropical trees, are less affected by variations in weather than are oil-yielding plants of the more temperate zones, though of course hurricanes, plant disease, and labour conditions have from time to time some effect upon the potential supply of their fruits. The olive tree, on the other hand, is much influenced by weather conditions and the yield may vary greatly from year to year in any given locality. But the olive yields a superlative oil. The weather also greatly influences the size of the crop of oil seeds from annual plants of the more temperate regions, especially flaxseed.

The weather even influences the production of animal fats, including milk fat, though its influence upon this is perhaps less obvious. The weather determines from season to season the amount of feed available upon pastures and ranges, and this determines the production of milk fat and the numbers and state of nutrition of beef cattle.

The corn crop is the most important single factor in lard production. The weather determines in large measure the size of the corn crop, and thus the weather is indirectly one of the factors that determine the magnitude of production of lard, of butterfat, and of tallow.

Cottonseed oil and corn oil stand in a peculiar position. Cottonseed is a by-product of cotton production. The volume of seed available depends therefore more upon the cotton crop than upon the price of cottonseed oil. But the cotton acreage depends upon the planter's estimate of the probable future price and supply of cotton. The actual production depends upon the acreage and the yield. The yield in turn depends upon the weather, and upon the boll weevil and other insect and fungous enemies of the cotton plant. The quantities of cottonseed available do not depend, therefore, upon the demand for and price of cottonseed oil.

The supply may be diminished independently of the cotton crop but not increased. If prices of oil and therefore of seed are very low, farmers may elect not to sell their seed to the oil crushers but to use it for fertilizer or feed. In former decades this was a more important factor than it has been since the United States has ceased to export any considerable quantities of cottonseed oil. The factors that in those days, and to a lesser extent even today, determined whether all or only a fraction of the available cottonseed were to be crushed, were not merely the price of the oil but also the price of fertilizer and of feed, the cake being used for both of these purposes. Cottonseed meal must be fed with caution, but remains a most valuable protein concentrate.

The production of corn oil is relatively small. It is perhaps less affected by the various influences above enumerated than any other fat, for it is a by-product of the corn-products industry and to a lesser extent of the milling of corn meal. The principal articles manufactured by the corn-products industry are corn starch, dextrine and other adhesives, glucose, so-called gluten feed, corn oil, and a large number of other products of lesser importance used directly or indirectly in the arts.

Starch and glucose and, to some extent, dextrine are used as foods, but a large part of the output of the industry is used in the arts. The demand for the majority of industrial corn products therefore tends to fluctuate with the business cycle. The raw material, corn, is always available in abundance for industrial uses, since the industry consumes an insignificant fraction of the crop, though at different prices because the crop varies greatly from year to year.

The consumption of corn meal also represents an insignificant portion of the crop. Moreover, from much of the corn ground into meal no oil is produced; the oil-containing fraction of the grain goes to feed. All these are conditions that warrant the conclusion that the quantity of corn from which oil is obtained as a by-product does not fluctuate greatly from year to year. So far as fluctuations do occur they depend primarily upon general business conditions and secondarily upon the price of corn. Hence, unlike cottonseed oil, the supply of corn oil does not fluctuate violently from year to year with the weather, although the trend of production has been upward in harmony with the general expansion of industry and population growth.

CONDITIONS AND TRENDS OF CONSUMPTION

In feeding stuffs very great quantities of fats disappear but, with a few unimportant exceptions, they always do so as a constituent of some element of the feed and not as such. Thus the amounts of fat that disappear with grain and legumes fed to animals are truly enormous. In food, as in feeding stuffs, very large quantities of fats are consumed incidentally in the ingestion of meats, fish, poultry, milk, and other animal products and in the form of fats and oils naturally contained in cereals, legumes, fruits, vegetables, and other plant products. However, in the human dietary fats and oils are also consumed practically as such, for example as butter, salad oils, and cooking fats.

QUANTITATIVE DATA UNSATISFACTORY

Not only because fats and oils disappear both in the arts and in the dietary in a vast variety of ways, but for other important reasons, the quantitative study of the consumption of fats and oils encounters great difficulties.

In the first place, the very concept of consumption is elusive. Presumably one should exclude, in spite of its potential significance, the fat content of large numbers of meat animals that perish from natural causes and from which no attempt is made to recover the fats. Commercial consumption is a significant concept, but it excludes a large amount of fats in products that do not enter into trade, and a large amount that are incidental components of meats and other products that do enter into trade, for example, tankage and garbage. Wastes in the process of recovery are considerable; hence the fat content of the product treated may be considerably greater than the fats recovered as such plus those incidentally retained. The amounts ingested by animals and by human beings would be worth knowing, but they would exclude large quantities that are wasted in various ways on the farms and in households and public eating places. Furthermore, a distinction may be drawn between intermediate and final consumption. Large amounts of fats are contained in cereals, nuts, milk, oil cake, and other products fed to animals; these are in part used up by the animals, and in part are converted or stored up and appear in dairy products or slaughtered animals. Hence a total of the fats consumed by animals and those derived from animals would contain, in effect, considerable duplication. The difficulties are enhanced by the fact that our statistical information is defective especially in respect to the data on animal fats.

For the United States, the number of animals killed in inspected slaughterhouses is known; the total slaughter at wholesale is reported by the census of manufactures; but only a guess can be made at the number of animals in other uninspected, principally rural, slaughter. Moreover, there is no clear-cut idea of the fat content per average animal slaughtered; the long-time trend seems to be downward, because younger and lighter animals are coming to slaughter in response to the increasing demand for lamb instead of for mutton, for bacon-type hogs instead of for the lard type, and for baby beef instead of for three- to four-year-old steers. To take even the best ratios derived from packing-house practice and apply them to the total wholesale slaughter of cattle, calves, sheep, and hogs as reported in the biennial censuses, and also to the farm and retail slaughter for which the data are officially admitted to be crude estimates, would be little better than intelligent guessing. At the best, there are gaps in the information so large as to make statistical conclusions on several important components, and on fats and oils as a whole, exceedingly unreliable.

It would be of interest to know, for each of the several sources of fats:

- The amount wasted for lack of any attempt at recovery;
- The amount ingested;
- The amount wasted in households and public eating places;
- The amount fed to animals; and
- The amount used in industrial production other than food.

We should like these data for fats produced, fats imported, and fats exported. We should like to distinguish between gross and net consumption. Unfortunately, for many of the sources of fats the statistical information is exceedingly defective and it is difficult to reach reliable bases upon which to make the estimates necessary for arriving at supplementary data. We have made attempts to reach a rough approximation to the amounts of fats and oils available for human consumption or industrial use in the United States in recent years, but at several points gaps are so wide and the procedure is so open to objection that we do not feel justified in presenting here even a preliminary approximation, lest in spite of reservations and qualifications it should be taken for more than it would be worth. In the discussion that follows, therefore, we have incorporated quantitative data only to a very limited extent.

The two principal items omitted, fat of dressed meats and milk fats (including butter), are so difficult to estimate that we have not included them; yet their sum probably exceeds the sum of the fats and oils here itemized. They go largely into edible uses, except as milk is fed to animals, though in considerable measure meat fats are wasted in the household or subsequently recovered from household wastes for industrial use. The principal other items omitted are fats incidentally consumed in fish, poultry, grains, vegetables, and fruits. Judging from estimates of Raymond Pearl for the period, the sum of these items is probably less than a billion pounds a year, exclusive of the fat in grain fed to cattle.

Table. Consumption of Certain Oils and Fats

Oil or Fat	Average Consumption 1921-25
Vegetable	
Cottonseed	1,074
Coconut	472
Palm kernel	13
Palm	90
Corn	102
Olive	104
Peanut	20
Linseed	654
Chinawood	76
Soy bean	17
Castor	35
Total	**2,657**
Animal	
Lard	1,552
Oleo oil, oleostearin, edible tallow	147
Inedible tallow	346
Other inedible animal fats	364
Total	**2,409**
Fish	
Fish oils	134
Grand total	**5,200**

FATS AND OILS IN DIET

The function of fats and oils in the diet is mainly to furnish energy to operate the animal machine. In the body, fats are burned as truly as though they were burned in a candle or under a steam boiler, and the end-products of the combustion are the same—carbon dioxide and water. Moreover, the amount of energy they furnish is the same, namely from 9 to 9.4 calories to the gram, whether they are burned within or outside an animal body. (A large calorie, which is the one here used, is the quantity of heat necessary to warm 1 kilogram of water from 0 deg Centigrade to 1 deg Centigrade.

A small calorie is the quantity of heat necessary to warm 1 gram of water from 0 deg Centigrade to 1 deg Centigrade.) They yield more energy than the other important classes of foodstuffs, such as carbohydrates and protein (albumin) Conditions and Trends of Production—Vegetable oils major products), which furnish from 3.8 to 4.2 calories to the gram.

Neither fats nor carbohydrates ingested are wholly burned at once unless they are needed for the operation of the animal machine. If not so needed, carbohydrates are for the most part completely changed in character by conversion into the fat characteristic of the species. This is stored in the body as a reserve against the possibility of a future period of food shortage. Indeed, most of the fat of domesticated animals is produced from starch.

Thus the lard of the hogs of the corn belt is mainly derived from the starch of corn (maize). It has the characteristics of fat normal to the animal, is normally stiff, and hence is especially prized. Unlike ingested carbohydrate, the fat of the food, if it is not at once burned, is changed comparatively little. For the most part it is deposited in but slightly modified form with other fat, made by the animal from carbohydrate or protein, in the storehouses for fat—the adipose tissue under the skin, in and about the viscera, and elsewhere. Therefore, if the food fat differs in its properties from the fat natural to the animal and if there is a great deal of it in the feed, the storage fat formed by the deposition of food fat gives an abnormal character to the fat which is obtained from the animal after slaughter.

The dietary fats may be divided into four groups:

1. Those consumed as such on the table;
2. Those employed in the preparation of foods;
3. Those consumed incidentally in the ingestion of meats and other animal products; and

4. Those consumed incidentally in cereals, legumes, fruits, and vegetables.

The fats consumed as such on the table in the United States are principally butter, butter substitutes (chiefly coconut and oleo oil), and salad oils (principally cottonseed, corn, and olive oils). The fats and oils used in the preparation of food are butter, lard, lard compounds (for the most part cottonseed oil), cottonseed oil as such, corn oil, peanut oil, butter substitutes, and, in certain types of confectionery, coconut oil and cacao butter. Fats and oils consumed incidentally in the ingestion of meats and dairy products are found principally in milk, cheese, poultry, eggs, beef, mutton and lamb, pork, and fish. The fat of meats is of course to a material extent not ingested but wasted except in so far as it is recovered in garbage and similar greases. The fats of the cereals, legumes, fruits, and vegetables as a class are important in the aggregate nutritionally, but have little direct commercial importance, since few foodstuffs are purchased on the basis of fat content. Practically all foodstuffs except sugar, water, and certain condiments contain some fat. Fresh fruits and vegetables contain merely traces; nuts, on the other hand, are rich in fats, as are also chocolate products. Oat meal and corn meal that is made without removing the germ are relatively rich in fat; wheat flour is poorer.

A study of the trends of consumption encounters the difficulty that statistics of fat in the diet over a series of decades are not available. Certain inferences seem nevertheless warranted. Trends of consumption may be grouped in two classes according as they are due primarily to changes in habits or primarily to substitutions made by manufacturers of which the consumer may or may not be aware. The two classes of course overlap, for in certain cases both factors play a part. Thus a manufacturer may create a new product which leads to a new food habit or, vice versa, consumer demand may lead to substitution or creation of a new product by the manufacturer.

INFLUENCE OF CHANGES IN FOOD HABITS

Change of food habits may be of two sorts. It may be purely

quantitative or it may be qualitative. By a quantitative change is meant increase or decrease in consumption of a given fat rather than the substitution in the diet of one fat for another. By qualitative change is meant primarily a substitution of one fat for another. However, since for physiological reasons the intake of food is practically constant for any individual under any given set of circumstances, a decrease in fat ingestion usually involves either a substitution for it of some other kind of food, or of some other kind of fat.

Conversely, an increase usually involves a decrease in consumption of some other kind of food or of fat. Since all fats and oils have very nearly the same food value, one can be substituted for another without change in the volume of food ingested. This statement applies to energy values. Certain fats contain small quantities of chemical substances of unknown chemical nature but of great importance to health. These are known as vitamins or food accessories.

Different fats contain different amounts of them depending upon their origin, method of preparation, and other factors. From the point of view of their vitamin content all fats are not of equal nutritive value; but these are considerations. However, if another food be substituted, the volume of the diet is thereby necessarily increased since no other food has, weight for weight, so great an energy value. It follows that if the replacement of fat in the diet by other foods goes too far the diet must become very bulky if it is still to furnish the same energy value, and this fact sets mechanical limits to the substitution of other foodstuffs for fat in the diet.

Since the caloric requirements of an individual, as stated above, are constant under given conditions, it follows that under these circumstances a reduction in consumption of a fat involves the substitution for it of either an equal amount of some other fat or a corresponding amount of some other foodstuffs. But conditions do not remain constant for any person. He grows old and requires less food. He may become stouter or leaner, then consuming either more or less food. He may change his means of livelihood from one requiring hard manual labour involving high food requirements to a sedentary occupation requiring a low food intake. He may migrate from a very cold climate where food requirements are relatively high to a hot one where they are relatively low. In any nation all these changes are taking place in some individuals in one direction, in others in the opposite direction. For short periods, the result is a reasonably constant consumption.

Over a long period of time this is no longer true. The age distribution of populations changes. If the birth-rate falls and the average span of life lengthens, the proportion of old people who consume little food increases, the proportion of young people who consume much decreases. If it becomes unfashionable to be stout, less food is consumed by the nation as a whole. If the proportion of manual labourers becomes less and that of machine tenders and sedentary workers greater, the per capita consumption of food tends to fall. Since fat is the most concentrated form of food energy, its intake tends to be reduced rather more than that of other forms of food.

In the United States exactly these changes have been taking place for decades; and there is evidence of a corresponding decline in the nutritional use of fats and oils, as part of a general reduction in the per capita food requirements. This reduction is the result of substitution of machine labour for man labour, the decline of outdoor employment in severe winter weather, the improved heating of buildings, the trend to lower average body weights of the people generally, and changes in the age distribution of the population. The population is acquiring more sedentary characteristics, needs less food, and the average per capita requirements of foodstuffs in terms of calories are automatically reduced. Most of these factors are the result of the economic evolution of the country and in essence reflect improvement in the standard of living.

The rise of the standard of living is responsible for certain other changes in food habits. It results in a diversification of the diet with increase in the use of dairy products, fruit, vegetables, and sugar, and decline in the ingestion of cereals and fats. This general statement does not hold for the fat of milk, the use of which is on the increase. Fat consumed incidentally to the ingestion of meats and dairy products has always been heavy. The consumption of milk and poultry fat is increasing, that of beef, sheep, and hog fats is declining. Plain cooking is being replaced by more fancy cooking, which includes more discriminating culinary uses of fats and oils. American practice has departed from the British custom of boiling vegetables without fat. Much of the crackers and bakers' bread consumed now contains shortening.

These facts might seem to suggest increased use of fats and oils. A survey of the entire field, however, suggests that the net result is a lower per capita ingestion of fats in general. The heavy fat rations of hard workers are a thing of the past—fat-backs and sow-belly are no longer staples. We use more butter, but less hog fat.

The current public taste for younger and lighter-weight animals represents a substantial reduction in fat. It seems fair to infer that, milk fat and poultry fat aside, the decline in the ingestion of fats applies to animal fats. Over a generation, indeed, it is possible that there may have been an absolute increase in the per capita consumption of vegetable oils with an absolute decline in the per capita consumption of animal fats outside of milk fat.

INFLUENCE OF CHANGES IN FOOD MANUFACTURE

The changes in trends of food-fat consumption are not all by any means due solely to change in habits. Some of those that are quantitatively extremely important are due to substitutions on the initiative of manufacturers of food-fat preparations—for example, the substitution of lard compounds for lard—or to the increasing availability of new types of fat—for example, coconut oil. Some of the possibilities of substitution in food uses have already been touched upon incidentally. However, since dietary consumption is the major and the higher or premium use of fats, the practice and possibility of substitution require elaboration because they are basic to any consideration of the commodity economics of this important group of raw materials.

The substitution of one fat for another, indeed the mere possibility of such substitution, naturally has the greatest influence upon prices.

Given sufficient price inducement, one fat may in some cases displace another with the widest repercussions upon the producer, the farmer, the trader, and the manufacturer. It is one of the purposes of the *Fat and Oil Studies of the Food Research Institute* to present studies from time to time upon far-reaching movements of this general character. That considerations of price, costs of conversion and refining, lack of technological skill, and legislation in the interests of public health and sanitation limit the volume of inedible fats turned to edible uses, has already been pointed out. But there are other factors as well that limit not merely the diversion of inedible fats to edible uses but also the substitution of one edible fat for another. The most prominent of these are greater or lesser adaptability of different fats to use in the preparation of different foodstuffs, and psychological and sentimental factors that in the food industries play a greater part than in the arts.

CONDITIONS AND TRENDS OF CONSUMPTION

DEVELOPMENT OF COMPOUNDS AND MARGARIN

The degree of adaptability of fats to different culinary practices is a result of their physical behaviours (consistency, flavour, etc.), not of their nutritive properties. It is the constant effort of technologists to widen the adaptability of fats to different uses. One of the more notable of their achievements is the production of lard compounds, as a result of which the use of lard seems relatively declining, that of lard substitutes increasing. And there has also been a pronounced decline in the use of tallows as cooking fats in the factory and the home. The principal use of lard is as a shortening and cooking fat. In the United States lard compounds are substituted for it in these uses to a considerable degree.

This means the substitution of beef stearin and cottonseed oil (hydrogenated as well as unmodified) for lard. This very extensive substitution has been possible partly because lard compound is a very good imitation of lard, partly because there is no adverse legislation, but also because sentimental considerations on the part of the consumer favour rather than hinder the substitution. Lard compounds usually lack the flavour of lard. They are for the most part quite bland, and this lack of distinctive flavour causes them to be preferred by a considerable portion of the population. Many brands of lard compound are made wholly from vegetable oil. They contain no animal fat whatever.

The manufacturers have been clever enough to feature this fact and thus to appeal to that portion of the population that assumes the vegetable fats to be purer, cleaner products than animal fats. Indeed, manufacturers have gone so far as not to represent their products as lard substitutes at all but as vegetable shortenings, sold commonly under their own distinctive brand names.

Manufacturers of margarin, the other notable achievement of technologists in the food-fat preparation field, have not been equally successful. This is due in part to psychological and esthetic considerations (taste, colour, spreading qualities). Against esthetic and sentimental factors successful substitution is difficult. Success demands that the substitute be a nearly perfect imitation. Margarin manufacturers have not always been wholly successful in making their product resemble butter exactly—no doubt in part due to adverse legislation. Within the industry itself there has been a drift from animal fats as raw materials to vegetable fats—especially coconut, cottonseed, and peanut oils. The use of peanut oil continues in the face of the competition of cottonseed oil because of the legal requirement that margarins represented to the consumer as being nut products must be made of nut oils.

SUBSTITUTION IN OTHER FOOD USES

Restrictive legislation, an important influence in margarin manufacture, is also completely preventing substitutions in other foodstuffs. The manufacture of filled milk is prohibited, and the manufacture of filled cheese has been taxed practically out of existence. Filled milk is skim milk in which coconut oil has been emulsified—in other words, it is milk in which the natural fat has been replaced by coconut oil. Filled cheese is made from skim milk and some fat foreign to milk or from partly skimmed milk and such fat.

Ice cream is another product in which expensive milk fat might be replaced by a cheaper one—for example, coconut oil. This substitution is not generally permitted by food-control officials, though it is reported that such a product is sometimes made for purely local consumption.

In the confectionery industry there are similar possibilities of substitution. Chocolate is made from the cacao bean by first roasting in revolving steel cylinders, after which the hulls are removed by machinery. Not to be confused with the coconut. The cacao bean is the seed of *Theobroma cacao,* L., a tree native to Central America but grown commercially in many places in the tropics. The seeds are borne in large pulpy fruits, each about 10 inches long and 4 inches thick and containing from 20 to 40 seeds.

The coconut, on the other hand, is a true nut produced by a palm, *Cocos nucifera,* L. The beans are then crushed and freed from the germs. The coarsely crushed product, freed from hulls and germs, is known as cacao nibs. The nibs, finally, are thoroughly ground between stones, the material being reduced to a thin paste which, on cooling, sets to a firm cake. It is known as unsweetened or plain chocolate. It contains over 50 per cent of cacao fat, which is known as cacao butter. It is produced from chocolate by expression, and is a soft solid much prized for certain uses in pharmacy. The press-cake that remains is known as cocoa or, when ground up, as powdered cocoa.

Many kinds of confections are coated with chocolate. They are made by dipping the center into a melted mixture of chocolate and sugar. If much sugar has to be added to the chocolate coating to give it the desired sweetness, it is necessary to add cacao butter in order that the melted mass may be sufficiently liquid. It would be perfectly feasible to substitute for the expensive cacao butter a cheaper fat such as coconut oil. The addition of such a fat with its higher melting point would, moreover, be advantageous because the candies would be less likely to become soft, sticky, and unsalable in warm weather.

It has been held that the use of a fat other than cacao butter is not permissible if the confections are to be sold as chocolates, for the consumer would be deceived into believing that he was receiving a product made solely from the cacao bean with, of course, sugar, etc. In the case of confections not so sold but offered for sale under some fancy distinctive name the use of coconut oil is permitted. These are mostly cheap products sold by the unit for a nickel or a dime. Coconut oil may also be used in frozen products of the type of Eskimo Pie. Owing to the need for using additional cacao butter in products sold as chocolates it often happens that when the price of cacao beans is very low cacao butter is produced without any cocoa as a by-product. The beans, hulls and all, are simply run through an expeller to extract the butter while the press-cake, which contains all the cocoa, is used as fuel or thrown away.

In short, the trends in the use of fats in dietary uses may be summarized thus: The per capita use of fat as a fuel, a staple, has declined. Fats and oils are used in more specialized states, more discriminatingly, in our present diversified diet. Substitutions are widely practiced, but they are less varied and diverse than in the arts, partly because more stringent demands are made by the consumer in regard to the adaptability of a given fat to a particular culinary use, partly because of esthetic and sentimental considerations, and finally because of legislative restraints.

FATS AND OILS IN THE ARTS

The preparation of fats for food uses is on the whole comparatively simple. The use of fats in the arts involves in many cases intricate, complicated, and multiple manufacturing processes. Many of these processes are secret—at least when first introduced. In consequence, in studying consumption and trends of consumption in industry all the difficulties of studying consumption of edible fats are encountered and in addition many others.

Changes are more rapid, as well as less easily recognized, than in the food field. Yet since many fats can be used either in food production or in the arts, the trends of consumption of one class of fats cannot be understood without knowledge concerning the other. In most years cottonseed oil is not an important ingredient of soap, but *foots* are used to a more important extent. Both commodities have diminished in importance during the period that coconut oil has been increasing. The principal reason cottonseed oil is not used in greater volume is that it brings a better price in the edible-oil market and consequently, unless there is ample surplus, the bulk of it goes to edible uses. Foots, on the contrary, can only be used for soap or be distilled for their fatty-acid content.

The various grades of greases are also used in soap; and the volume so used may have increased materially, perhaps keeping pace with the rapidly increasing production of grease and the striking recent development of the rendering industry. Ordinarily coconut oil is more expensive than the other fats used in considerable volume in soap. Its special characteristics give it a premium over other materials. Tallow, white grease, and palm oil are ordinarily about the same price, although palm oil is apt to be cheaper than the other two and white grease at times is far higher, probably because it is now and then wanted in European markets as a substitute or adulterant for lard.

Economies have been introduced in the use of oils in tin plating and in the manufacture of woolen cloth. Undoubtedly, the use of vegetable oils as illuminants is declining to the vanishing point. The vegetable or animal-oil lamp and the tallow candle have ceased to exist in households, except for special occasions. There is, of course, a small persisting use of oils as illuminants for sacramental purposes, and there is quite a material use of fats for stearic-acid candles. To some extent, stearic acid is being replaced by paraffin. With the decline in number of work animals has come reduction in need for harness dressings. The relative use of fats and oils as belt dressing has probably declined, though the absolute quantity may have increased. In pharmacy, with the exception of cacao butter, fats have been replaced in ointments and salves by petrolatum and lanolin (wool grease).

The position of fats and oils as lubricants is difficult to determine. The amount of lubricants used has, of course, expanded enormously; but in this expansion animal and vegetable fats and oils have shared relatively much less than lubricants derived from petroleum. The absolute quantities of animal and vegetable fats and oils used in lubricants may not, however, have been reduced (indeed, an increase may have occurred), even though their proportion in the total mass of lubricants has been substantially lowered. Large amounts of animal and vegetable fats and oils are being used mixed with mineral oils in the preparation of special lubricants.

In estimating the trend of the use of animal and vegetable fats and oils in paints, difficulties are again encountered. The proportion of wooden buildings, requiring exterior painting, is declining relative to buildings constructed of other materials and requiring little exterior painting. For many of the coarser and cheaper paintings mineral substances are being widely used. In the treatment of floors, waxes are rapidly replacing paints. For interior household painting, and for high-grade painting and varnishing in general, the drying oils held their place until the advent of plastics derived from cellulose. These have practically replaced paint and varnish on new automobiles and are making heavy inroads with furniture, railway cars, refrigerators, and for interior uses in rooms such as kitchens and bathrooms.

SPECIAL DEMAND PROPERTIES

The several fats and oils have many properties in common; but some of them possess peculiar qualities. It follows that they are both complementary and substitutable. When a particular fat has unique properties, the commercial use will be inelastic to some extent. When a particular fat has only properties common to them all, the commercial use is elastic. In so far as fats have peculiar properties, there is a range of uses within which they are not substitutable, where competition between them is subordinate.

In the range within which fats and oils have properties in common and are substitutable, competition is predominant. Fats possessing peculiar properties carry a premium, within limits; fats possessing only common properties compete with each other on a close margin. The long list of commercial fats and oils represents a series of overlapping raw materials, with ascending prices in proportion to their peculiar properties or particular desirabilities. The lowest reclaimed grease stands at the bottom; at the top stands unsalted butter. The technologist is continuously endeavoring to enlarge the list and range of substitutable uses and to narrow the list of premium uses.

A few illustrations will make the commercial situation clear. The outstanding drying fats are linseed and tung oils, and for certain of the most particular paint and varnish uses they stand alone. But in the manufacture of common paints, it is practicable to add a proportion of semi-drying oils such as soy bean and fish oil. For the finest paints and varnishes the demand for linseed and tung oils is inelastic; for the commoner paints, the demand is elastic.

There is little price competition from the other fats at the top, but sharp price competition at the bottom. Linseed oil is an edible oil, but it is too valuable as a drying oil to serve as an edible oil in the United States, quality considered. The drying oils cannot be used as lubricants.

Candles are made from stearic acid. In the manufacture of candles, therefore, fats containing a high proportion of tristearin must be sought out, or else triolein must be converted into tristearin through hydrogenation, which is usually too expensive to be practicable. This tends to give a certain priority in this use to fats containing a high proportion of stearic acid, and such fats are commonly purchased on the basis of the titre test, by which the content of stearic acid is judged. This applies particularly to superfine hard candles; for the commoner candles, softer fats and paraffins are substitutable.

SUBSTITUTABILITY IN SOAP MAKING

Tallow is regarded as one of the fats most desirable for soap because it yields a soap that is both hard and of good colour. It has gradually come to be displaced to a certain extent by other fats so that at present little if any soap is produced from tallow alone. It is practically always blended with other fats, partly because of price considerations, but perhaps even more largely because better soap can be made from such blends than from tallow alone. While tallow soap gives a good and lasting lather, it dissolves rather slowly and so lathers slowly. This is remedied by adding other fats which yield soaps that dissolve more quickly and so lather more easily.

A number of oils are used for this purpose, but especially coconut oil, which has the further advantage of retaining a large proportion of water in the soap, thereby increasing the soap makers' yields. Soap containing coconut oil is more soluble in salt water than most other soaps except palm kernel oil soap. It therefore lathers in sea water; hence coconut oil is used in so-called marine soaps. Palm kernel oil soap has properties very similar to those of coconut oil soap and may be used for the same purposes. In the United States this oil is relatively little used, but in Europe it is very largely employed interchangeably with coconut oil. Coconut oil, moreover, is especially valuable in soap chips because its soap, while hard and brittle, is very soluble.

The washing machine has created a wide demand for such soap chips. Coconut and palm kernel oils occupy a premier position in the manufacture of marine soaps and laundry chips. For these uses other fats cannot be substituted for them, though both oils can be substituted for other fats over a wide range of other uses. There is a wide range of adaptation in the manufacture of toilet, laundry, and household soaps. In these, a number of animal fats and vegetable oils are substitutable, under technological procedures, largely on the basis of price. Manufacturers of trade-mark soaps endeavor to keep them uniform in order to retain established markets. There is a certain range within which they can substitute the raw materials and still maintain uniformity in the soap, and within this range they purchase the raw materials on the basis of price.

Outside of this range they cannot substitute fats with maintenance of uniformity, but must buy certain fats at going prices. The properties of trade-mark toilet soaps are held more rigid and invariable than those of trade-mark household and laundry soaps. The properties of trade-mark household and laundry soaps are held more rigid than in the case of soaps not carrying trade-marks. The lower the usage, the less the necessity for uniformity. Under these circumstances, the range of substitution varies inversely with the price of a soap. Since common soaps are manufactured on a price basis, soap makers under efficient technological practices draw their raw material from every conceivable vegetable and animal source. Indeed, they sometimes substitute other raw materials for fats and oils.

Thus in yellow laundry soap there is a good deal of rosin if the price warrants. Rosin acids form sodium salts which are freely soluble in warm water, lather well, and have good cleansing action. Rosin, however, is not used in white soaps; it makes them not merely yellow but also sticky. Hence also, it is not used in soap chips. Besides rosin, a variety of other substances are used at times, some of which are merely fillers or adulterants. The commonest is sodium silicate, familiarly known as water glass, of which as much as 20 per cent of the weight of the soap is sometimes used. Opinions differ as to whether this is to be regarded merely as a filler and make-weight or as a useful ingredient. Indeed, the substitution of other materials for fats and oils in many directions has been the endeavor of chemists for some time with considerable success, especially in the substitution of mineral products for fats and oils in lubricants, and of new types of varnishes and lacquers, free from vegetable oil, for the older types in which drying oils serve as the medium. Of course, such substitutions are not possible in the edible field, in which substitution is limited to one fat for another.

The uses of coconut oil are of interest in this connection. It is widely used in soap, especially marine soap, and in margarin.

It is, however, a poor lubricant, and compounds containing a large proportion foam badly when used for deep frying. Coconut oil is therefore a premium material for marine soap, an available material for common soap, but not to any important degree available for lard compounds or for lubricants.

Taking the situation as a whole, about all that can be said in the absence of more adequate statistics is that the trend of per capita consumption of fats and oils is probably on the increase, owing largely to increasing consumption of soap. Probably per capita consumption for food is somewhat decreasing. In industry, aside from soap, per capita use is possibly decreasing because of substitution of petroleum and synthetic products, though it is not improbable that the total amount of fats and oils consumed in industry outside of soap factories is increasing due to population growth. What the exact quantitative relations are, it is at present impossible to estimate.

INTERNATIONAL TRADE IN FATS AND OILS

FATS and oils Figure heavily in international trade; in particular the volume of trade in oil nuts, oilseeds, and vegetable oils has increased with the exploitation of tropical sources of supply, the development of technology, and the relative decline in animal fats.

Although statistics of international trade in fats and oils are less inadequate than statistics of production and consumption, they are far from satisfactory and present so many difficulties that no summary presentation seems desirable at present. In subsequent studies of particular commodities the current position and the trends can be set forth in some detail with appropriate qualifications and reservations.

INTERNATIONAL POSITION

The international trade in fats and oils flows largely in a few prominent streams. Palm and palm kernel oils come chiefly from Africa, though the Dutch East Indies are beginning to enter upon this trade. Practically all soy bean oil comes from Asia. Peanut, rapeseed, and cottonseed oils come from several regions.

Coconut oil comes largely from the Philippines, the Dutch and British East Indies, and Oceania. Large amounts of cattle and sheep fats come from Australasia and Argentina. The United States and the lower Danubian area are practically the sole exporters of lard. Imports are much less specialized than exports, though many countries import predominatingly more of one fat or oil than of others, depending on political affiliations. By and large, the more backward regions are prominent as exporters, while the highly civilized regions are prominent as importers. This is due partly to fundamental conditions affecting production, partly to considerations affecting consumption. But many countries export some fats and oils and import others, so that their net position in international trade in fats and oils as a whole may be very different from their position with respect to any one.

The term self-sufficiency, frequently used in this connection, may mean something or little, or may be actually misleading. It is technically correct, for example, but essentially misleading to say that the United States is not self-sufficient in cotton. We import each year some 200,000–500,000 bales, but our annual exports average well over 6 million bales. Cotton is not a unity. The imports consist largely of certain kinds that the United States does not produce at all. The cotton imported could, however, be produced, but the cost would be excessive. In view of the manifold sources and uses of fats and oils, it might similarly be misleading to say that any country, merely because it imports fats and oils, is not self-sufficient in these products.

In the analysis of the international position of a country, the position as net exporter or net importer may or may not be significant. To be a net importer may signify a shortage, irrespective of price; thus the United States is a net importer of nickel because it has no deposits. But a country may be a net importer solely on the basis of price; the United States is a net importer of paper, not because it lacks the resources in its forests, but because at present the forests of Canada furnish a cheaper product. A country may be a net exporter because of the position of a primary industry; in this sense, the United States is a net exporter of copper. But a country may be a net exporter for incidental reasons; thus the United States is a net exporter of caustic soda.

For some countries it is difficult or impossible statistically to determine whether they are net importers or net exporters of fats and oils. Also, if this be ascertained, it may be difficult or impossible to determine whether the position is inherently necessary or merely incidental. A determination of the net position of a country in respect to fats and oils is particularly difficult because of the complexity of the situation, one manifestation of which is the inadequacy of statistics. Fats and oils pass in and go out not only as such, but under other names and in other forms.

The United States both imports and exports fats and oils for use in the manufacture of soaps; it also both imports and exports soaps. Lard is counted as an export of fat, while cured-pork products are not; yet backs, sides, and bacon are three-fourths fat, and hams and shoulders are one-third fat. A state may be a fat-importing country when considered from the standpoint of industrial fats, and a fat-exporting country when considered from the standpoint of dietary fats; this is the position of the United States. It is, therefore, likely to be misleading to judge of the position of a country in respect to its total uses of fats and oils solely by such a classification as net importer or net exporter as might be determined from the use of trade statistics. Let us consider the circumstances in different countries.

POSITION IN COUNTRIES AND REGIONS

Russia, before the war, was a heavy exporter of butter, linseed oil, and edible vegetable oils. The customary diet of Russia was poor in fats and meats, consisting predominantly of cereals and vegetables (mostly rye, cabbage, and beets). Because of a low standard of living, the industrial uses of oils and fats were limited. On the basis of the trade Figures, Russia was self-sufficient; but she was so only because the uses of fats and oils in that country occupied a position so low, in contrast with that of western European countries, as to seem abnormal. The position of Russia is just the opposite from that of the United States. The two countries are not incomparable in size, population, and agricultural resources. Russia could not be called a superior fat producer and the United States an inferior fat producer.

But Russia, in contrast with America, has a low standard of living, in respect to both dietary and industrial uses of fats and oils. Russia is a net exporter of fats and oils, but could easily consume what she exports if she were not a backward country; the United States is a net importer of fats and oils, but could easily produce more if she chose. Europe as a continent, outside of Russia, is a heavy net importer of fats and oils, although a few countries, such as Denmark, are net exporters. Oilseeds are imported both for the protein and the oil fractions. The continent is also a heavy net importer of meats and animal fats, though Denmark and the lower Danubian area are net exporters of animal fats.

The population of Europe is dense in respect to the agricultural area; and most countries are highly industrialized. Cereals, meats, animal fats and oils, vegetable fats and oils, must be imported. More could be produced of any one of these; the amounts that are domestically produced at any time largely determine the amounts to be imported. The adjustment between the different types of agriculture and the different imports of foodstuffs and industrial raw materials is mainly the expression of current efficiencies in agricultural technique, within the limits set by climatic factors and character of soil.

The United States is a net exporter of animal fats. It is a large net exporter of hog fats, both in the state of lard and in cured-pork products. In much smaller but still considerable amounts, it is a net exporter of beef fats in various forms. On the other hand, the import of live cattle is fairly extensive; and imports of milk fat, chiefly as butter, generally exceed exports.

The United States is a substantial net importer of vegetable oils. Cottonseed oil is the only one of which the exports are appreciable, and most of the exports of other vegetable oils are either small in amount, or are produced from imported materials, or are re-exports of oils of foreign origin. Practically all kinds of vegetable oils are imported, either as such or, much more largely, in the form of raw material. Imports of linseed oil (chiefly in flaxseed) and coconut oil (chiefly in copra) are the largest items, but imports of olive, palm, palm kernel, chinawood, and castor oil are also important, and several other kinds are by no means negligible.

The picture is heterogeneous. If all the imports and exports were lumped, the United States would be shown to be a net importer, to a small extent per capita. But that the country is a net importer does not mean that there is a shortage, in the sense of inability to supply needs from domestic production. It merely means that a small fraction of our heavy fat and oil requirements is imported because it is cheaper to secure it thus than to produce it. If one separates dietary from industrial uses, the country is a net exporter of dietary fats and oils and a net importer of industrial fats and oils. These are the relations as they actually exist, not as they necessarily must be or must continue to be.

COLONIAL SOURCES OF FATS AND OILS

To a large extent the oilseeds and nuts of commerce move to countries situated in the temperate zones from colonies and dependencies in the tropics. The colonies or dependencies of Great Britain, Holland, France, Belgium, and the United States are the largest sources of supply of tropical oilseeds.

During the early years of this century Great Britain had by far the largest trade of any European country in foreign oilseeds and nuts, and was the principal user of imported vegetable oil materials. As the years passed, however, Germany became a keener and stronger competitor in both the producing and the distributing branches of the trade. Her colonization activities are well known.

In several of the colonies fatty vegetable materials were the principal or among the principal articles produced, and great efforts were made to stimulate their production. In addition, steamship companies were subsidized in the colonial trade and German ports were encouraged to organize for the handling of oleaginous materials so as to compete with such well-established trading centers as Marseilles and Hull. Consequently, by 1913, German imports closely approached those of Great Britain; indeed, the British trade was becoming seriously exercised over loss of business.

France also had a large trade in vegetable nuts and oils before the war, but it did not reach the proportions of the British or the German. Because of her large domestic supply of olives, she was not dependent on foreign sources to the same degree. Shortly before the war, Germany far outstripped either Great Britain or France in her use of coconut and palm kernel oil. These, aside from linseed, were the principal vegetable fats imported. Great Britain, on the other hand, used a great deal of cottonseed, while France relied principally upon peanuts. Great Britain was the only one of the three that imported any appreciable amount of these fats in condition suitable for consumption. She imported considerable margarin and lard compound from the continent.

The physical and chemical characteristics of the oils determine their use in main part, but since considerable substitution can be made the usage differs somewhat from country to country according to the availability of the various oils. In France, for instance, peanut oil is doubtless more generally used for margarin and salad oils than elsewhere. Similarly in Great Britain palm kernel oil serves many of the purposes to which coconut oil is put in America and Germany. The availability of palm oil for soap manufacture in Great Britain doubtless permits a larger use of animal fats and coconut oil for edible purpose there than in this country. Per capita consumption of margarin is considerably smaller in Great Britain than in Germany and the Scandinavian countries, but her total takings of vegetable fats and oils are larger than those of any other European country, apparently indicating very large industrial usage of inedible products. Careful investigation would show differing practices from country to country, manufacturing technique having been adapted to the supply of raw materials.

COMPETITION OF DOMESTIC AND IMPORTED SUPPLIES

Every country with colonies or dependencies has more or less effective incentives, political and commercial, for increasing the trade of these colonies or dependencies with the mother country and with other countries. The mother countries accept oilseeds in return for services, interest charges, and manufactures. In such countries the colonial vegetable oils are substitutable to some extent for native animal and vegetable fats and oils. The animal husbandries of the United States, Great Britain, Holland, Belgium, and France must market their fats and oils in direct competition with certain vegetable oils obtained from the colonies. The price reactions are sometimes direct, sometimes indirect. The circuitous nature of some of these relations is well illustrated in the case of Denmark. Denmark imports oilseeds, expresses the vegetable oil, feeds the cake as protein concentrate to dairy cows, churns the butter from the milk and ships it largely into export trade, uses the buttermilk as a basis for a hog industry of which the bacon passes largely into export, and of the expressed vegetable oil makes butter substitutes (plus some imported margarin) to be used in the domestic market to replace the butter and bacon exported. The net result is a profit which forms a material proportion of the national income.

North of the latitude of the Alps lard is preferred in cooking; southward, vegetable oils are preferred. Germany and the other countries of northwestern Europe (except Denmark, a net exporter, because of the peculiar development of her agriculture) import hog fat from the two exporting areas. The lower Danubian area (a net fat and oil exporter) exports hog fat because it produces fat in excess of its customary dietary needs. The United States (a net fat and oil importer) exports lard because a large fraction, possibly a majority, of our people prefer vegetable lard substitutes. Hog producers in northern Europe endeavor by tariffs to make defence against imported lard. What would happen directly to the hog raisers of northern Europe, and indirectly to those of the United States, if it should come to pass that lard substitutes should supplant lard in Europe, as has been the case in the United States, is an interesting subject for speculation.

Not merely do the hog producers of northern Europe strive for tariff protection against imported lard, but the dairymen agitate for tariff protection against imported vegetable oils because these are the principal raw materials for the enormous butter-substitute industry of western Europe. Similarly, there is agitation in Mediterranean countries for tariff protection against imported oils even though these countries are heavy exporters of olive oil, because the import of other edible vegetable oils reduces the domestic price of olive oil. In character and degree this demand for protection differs in no essential way from that in other agricultural produce or in manufactures. If all the countries from which vegetable oils are derived were independent, rather than colonial and dependent, their oil seeds would obtain entrance to the importing countries less readily than they do now.

Another aspect of this competition deserves consideration in its bearing on agricultural profits and tariff policies. The prime products of animal husbandry are milk and meat; the carcass fats, as we have seen, are secondary products. Cheap imported vegetable oils, largely from the tropics, are substitutable with indigenous fats and oils over a wide variety of uses and thus make for lower prices of the indigenous fats and oils. The lower the price level of the fats, the greater the load to be carried by the meats or by non-fat by-products. Since the prices of meats are relatively high and advances are met by restriction of consumption, and since rising hide and by-product prices arouse competition of substitutes, low prices for fat obtained by packers and merchants tend to be reflected back to agriculturists in the form of lower prices for live animals.

So far as fat is concerned, the effect is greatest in the case of hogs, since the lard and the cured fat parts of the average hog exceed one-fifth of the live weight; in the case of the average steer, the net yield of commercial fat is relatively small, consequently the effect of low prices of fats on the price of the live animal is smaller. Competition of artificial leather, bone substitutes, and other substitutes for animal by-products operates in the same way.

In some measure these influences affect the remunerativeness of the entire system of diversified agriculture. Tariffs on importations of fats and oils, in so far as they might tend to raise the general price level of fats, would thereby tend to raise the return on the carcass fat produced in animal husbandry, to raise slightly in consequence the prices of live animals, and to improve slightly the profitableness of diversified agriculture.

With the exception of olive oil the major fats produced in western Europe are the product of animal husbandry; and production by no means suffices to meet requirements. Indeed, much of the animal fat production, including milk fat, is dependent upon the importation of oil seeds. The oil cake remaining after the expression of the oil is fed to cattle and hogs, thereby helping to produce fat.

The manure resulting from this feeding adds to the fertility of the soil. Thus the importation of oilseeds has three effects of the greatest value to the importing country. First, it contributes dietary and industrial fats directly. Secondly, it contributes concentrated animal feed by means of which meat and animal fats are produced. Thirdly and lastly, the feeding of the oil cakes produces very rich manure, with which heavier crops of grains, roots, and legumes are raised than would be possible otherwise.

The vital importance of the importation of oilseeds into western Europe was brought out during the Great War. When the allies stopped the importation of oilseeds and of oil cakes into the Central Empires, importation of these commodities into Scandinavian countries and into Holland increased by an amount not very far from the average pre-war importation into Germany. These countries became practically feeding yards for Germany. When the importation of oilseeds and oil cake into neutral neighbouring countries was brought under control by the allies, Germany began to feel a shortage of fat. This shortage was greater than was represented by her pre-war normal importation of fat, for inability to secure oilseeds and oil cake contributed to the deficiency of feeding stuffs. This necessitated a reduction of the cattle and hog population with a consequent reduction of the meat and animal-fat production. Because of the lack of such concentrated feeds as oil cakes the fat production—and especially the butterfat production—by the surviving animals fell below normal.

And this was not all. The lack of concentrated feeding stuffs reduced the volume and the fertilizing value of manure, which in turn resulted in smaller yields of cereals, legumes, roots, and other crops. It is obvious, then, that the oilseed trade is of vital importance to western Europe. It is not to be ignored by these nations in shaping their respective national agricultural policies.

CONCLUDING OBSERVATIONS

The presentation is a failure if it has not made clear that there has been little systematic study of the subject, that it presents matter of great interest to students of economic theory, that there are wide gaps in basic information, and finally that there are many problems important to American national welfare. Let us return for a moment to consideration of some of them.

We need to know the effect, prospective as well as present, upon the mixed farming which is so important to our general national well-being. Because until recently we have been net exporters of fats and oils and because we are still heavy exporters of lard, we have paid little attention to these questions. Since the war the situation has been changing and it is now high time that it be analysed objectively and taken into consideration in shaping national policy with reference to agriculture.

However, to appraise the situation touched upon in the preceding paragraph we need to know vastly more than we do concerning tropical and other fats that are offered for import into the United States—the coconut, peanut, palm, palm kernel, soy bean, and other oils. We need to know more concerning their costs of production, existing acreage, trends of acreage, outlook for increased or decreased production, probable demand in the several consuming countries, character of the world trade, and the like.

Comprehensive data and interpretation of data are needed for each case. In our own country we need to know many things of which we are now ignorant but which are none the less of great importance for our national welfare. We need comprehensive detailed statistics, now largely lacking, of the ways and forms of consumption of the several sorts of fats and oils that disappear each year. Only by studying such data for a series of years will it be possible to forecast with any reasonable degree of probability the trends of consumption and the probable future demand for the different types of fatty raw materials.

But such knowledge is essential if our national policy with reference to agriculture as well as to foreign trade is to be based upon a rational foundation rather than upon the accidental success before the Congress of one or the other of two warring groups representing supposedly or actually opposed economic interests.

These are but a few of the important questions that obtrude themselves promptly upon the student. They are not so much individual projects for research as broad general fields for investigation. They are so broad that much of the field can best be cultivated by government agencies—indeed, some of the problems cannot be attacked with hope of success in any other way.

No other agencies have the resources or the power to gather the necessary basic statistical data. Only when these have been collected, tabulated, and made generally available can other agencies undertake the study of many of the economic aspects of the fats and oils situation.

In the meanwhile it seems possible to mark off from these broader fields of investigation many specific smaller territories in which even now research may profitably be undertaken by agencies with lesser resources and less power than governments but with greater freedom of action and of opinion. It is towards the study and solution of problems of this sort concerned with the role of fats and oils in the world's economic situation that the Food Research Institute hopes from time to time to make some contributions.

2

Vegetable Crops

INTRODUCTION

The plant is the most important agent in crop production. Soils, cultivation, fertilizers, irrigation, and other factors, in a sense, are all more or less subsidiary. Soils are modified by cultivation, by adding manure or other fertilizers, by drainage or irrigation, and in other ways with the express purpose of changing the environment so as to stimulate plants to increased productivity. Hence, it is not surprising that from time immemorial extended observations and, later, experiments have been made upon the aerial growth of crops under varying conditions.

In fact an almost bewildering array of literature has resulted. But quite the converse is true of the underground parts. The root development of vegetable crops has received relatively little attention, and indeed accurate information is rarely to be found. The roots of plants are the least known, least understood, and least appreciated part of the plant. This is undoubtedly due to the fact that they are effectually hidden from sight. Notwithstanding the extreme difficulty and tediousness of laying the roots bare for study, it is not only remarkable but also extremely unfortunate that such investigations have been so long neglected.

It clearly seems that a thorough understanding of the activities of plants both aboveground and belowground and the ways in which these activities are favourably or unfavourably modified by various cultural practices should be basic for scientific crop production. Yet almost countless field experiments, selections, breeding, and testing of varieties, etc. have been carried on in all parts of the world with little or no knowledge as to the behaviour of that very essential portion of the plant, the absorbing system.

Similarly, in the study of soils, the greatest attention is given to the problems of the physics, chemistry, and bacteriology of this substratum and rather largely to the cultivated portion of the surface only. The soil, barring the living organisms which it supports, is perhaps the most complex, the most interesting, and the most wonderful thing in nature. Surely it should receive thorough investigation. But a study of the soil and the way in which its various relations affect yield without a consideration of the essential, intermediary absorbing system is more or less empirical. A complete, scientific understanding of the soils-crops relations cannot be attained until the mechanism by which the soil and the plant are brought into favourable relationships, the root system, is also understood.

The student of plant production should have a vivid, mental picture of the plant as a whole. It is just as much of a biological unit as is an animal. The animal is visible as an entity and behaves as one. If any part is injured, reactions and disturbance of the whole organism are expected. But in the plant, our mental conception is blurred by the fact that one of the most important structures is underground. Nor is the plant usually treated as an entity; it is often mutilated by pruning, cutting, and injuring the root system, frequently without much regard to the effect upon the remaining portion.

MODIFYING THE ROOT ENVIRONMENT

In both field and garden the part of the plant environment that lies beneath the surface of the soil is more under the control of the plant grower than is the part which lies above. He can do relatively little towards changing the composition, temperature, or humidity of the air, or the amount of light. But much may be done by proper cultivation, fertilizing, irrigating, draining, etc. to influence the structure, fertility, aeration, and temperature of the soil.

Thus, a thorough understanding of the roots of plants and the ways in which they are affected by the properties of the soil in which they grow is of the utmost practical importance. Before other than an empirical answer can be given to the questions as to what are the best methods of preparing the land for any crop, the type of cultivation to be employed, the best time or method of applying fertilizers, the application and amount of water of irrigation, kind of crop rotations, and many other questions, something must be known of the character and activities of the roots that absorb water and nutrients for the plant and the position that they occupy in the soil.

ADAPTATION OF ROOTS TO ENVIRONMENT

By more or less profound modifications of their root system, many plants become adapted to different soil environments, others are much less susceptible to change. Among forest trees, for example, the initial or juvenile root system of each species follows a fixed course of development and maintains a characteristic form for a rather definite period of time following germination. The tendency to change when subjected to different external conditions becomes more pronounced as the seedlings become older. But some species thus subjected exhibit much earlier tendencies to change than others and a widely different degree of flexibility is also shown. Hence, certain species, *e.g.*, red maple (*Acer rubrum*), because of the great plasticity of their root systems, are able to survive, at least for a time, in various situations from swamps to dry uplands.

Native herbaceous species usually show great plasticity of root habit, successfully adapting themselves to considerable differences in soil environment.

The plant on the left was excavated in the Great Plains of Colorado. Maximum branching occurred in the third and fourth foot, although the taproot reached a depth of more than 7 feet. The plant on the right, with a shallower but much more finely branched root system, was excavated from a "half-gravel slide" in the Rocky Mountains of Colorado.

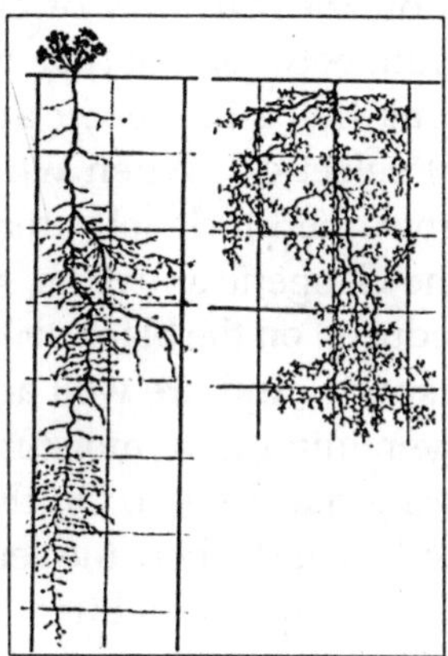

Fig. 2.1 A spurge (*Euphorbia Montana*) Showing Differences in Root Habit Resulting from Environment.

The wide range in soil and agricultural conditions under which vegetable crops are grown renders them particularly suitable both for the investigations of their root habits and also for a study of the agricultural significance of the differences encountered.

ROOT ADAPTATION AND CROP PRODUCTION

Enough work has been done to show clearly that among garden crops root adaptations frequently occur. These will be pointed out in the following chapters, although a mere beginning has been made. The field is enormously large and difficult but one rich in possibilities. It seems entirely probable that some of the best-yielding crops are able to outstrip others largely because of their greater efficiency in securing a greater and more constant supply of water and nutrients. On the other hand, the failure of a crop to thrive in a particular soil may be due to a lack of adaptation of its root system to the environment imposed upon it. Both of these conditions are illustrated by the growth of flax in India.

The agriculture in many of these areas is ancient, there have been few innovations, and the soil conditions have had time to impress themselves on the varieties of crops cultivated. A condition of equilibrium between the type of plant and the soil has been obtained, as there has been ample time for the operation of natural selection. When we compare the root system of linseed from the black-soil areas with that of the varieties grown on the Gangetic alluvium, striking differences appear. The roots produced on the black soils (of the Peninsula) are deep, somewhat sparse, and are well adapted to ripen the plant quickly with the minimum of moisture.

The type of gearing fits the soil. On the alluvium, where moisture is more abundant and where the aeration of the subsoil is poor, the root system is superficial but at the same time well developed. On the intermediate types of soil linseed produces a type of root about halfway between that of the black soils and of the alluvium. Once more the root system is found to fit the soil type. Further when we grow side by side on the alluvium these three classes of linseed, there is little or no adaptation of the roots to the new conditions, but the three types behave very much as they would in their native habitat.

The deep, sparse root system of the black soil areas is developed in the alluvium, although it is fatal to the well-being of the crop. When, the experiment is reversed and the types which suit the alluvium are grown on the black soils, there is again little or no adaptation to fit the new conditions. The linseed crop consists of a large number of varieties which differ from one another in all sorts of characters, including the extent and distribution of the roots. The root systems of the varieties are just as characteristic and just as fixed as the differences in the seed and other aboveground characters of these plants. A similar state of affairs obtains in other crops like wheat... and the opium poppy, and is probably universal all over India. Thus the differences between the root systems of varieties have been clearly shown and the economic significance illustrated.

Vegetable growing is an important phase of agriculture and one that is increasing at a rapid rate. In the United States practically every kind of soil is used for growing vegetables. Among the dozens or sometimes hundreds of varieties of the various vegetable crops, some undoubtedly have root systems more suited to a given soil environment than others. If the highest possible production is desired, attention should be given to selecting the variety that is not only climatically adapted but also best fitted to a particular soil, modified as favourably as lies within the power of the grower. It seems certain that a large part of the success in connection with the improvement of varieties consists in better adapting the absorbing system to the soil. As stated by one who has extensively worked on plant selection, breeding, root investigations, and other phases of agronomic science, "The more I learn about cultivated plants the more I am convinced that the future lies in the root-soils relation and in matters which influence it." An intimate knowledge of the habits of growth of the root systems of vegetable crops will enable the grower to space plants to better advantage. It should also permit him to intercrop or grow in succession such crops or mixtures that the soil volume will have a better distribution of roots and thus permit of methods of more intensive cultivation. Similarly, by means of proper crop rotations and occasionally cultivating very deeply rooting crops, the subsoil may be kept in good condition and the effects of drought mitigated.

INTERRELATIONS OF PLANT, SOIL, AND CLIMATE

In considering the importance of root relations in crop production, it should be clearly kept in mind that the plant, the soil, and the climate form a closely interlocking system of which no part should be overlooked or overemphasized. It is now rather generally recognized that climate and vegetation are the most important factors determining the character of the mature soil. 148 "The features assumed by the soil in its development from infancy, through youth, maturity, and old age, vary with the environment, especially with the climate and the vegetation." The effect of both climate and soil on the growth of aboveground plant parts has long been known. It has only recently been clearly demonstrated that the environmental factors which affect the root are not only those of the soil immediately about it but also those affecting the shoot which is rightly a part of the complex. Through the shoot the root system is influenced by the aerial environment. 23,103 The amount of light or the degree of humidity, temperature, etc. and the effect of these upon food manufacture, water loss, and other activities affect root development. In fact there is a rather close correlation between shoot and root development.

The complex relationship of plant, soil, and climate may be further illustrated in the use of fertilizers. They modify the habit of growth as well as the composition of the plant. For example, phosphates, when applied to soil upon which wheat or certain root crops are grown, promote deeper root penetration. This results in a greater water and nutrient supply for the plant. Earlier development and ripening may be promoted or drought mitigated. Soils thus fertilized will produce crops under an environment perhaps otherwise quite unfavourable. Vegetable production should be studied from the point of view of how roots and shoots of plants grow, and use should be made of the plant itself for indicating the direction of future research. An adequate discussion of the environment of roots (the soil), how roots are built to perform their work, and root habits in relation to crop production has been so recently given in "Root Development of Field Crops" that further statement seems unnecessary. For a general discussion of the effects of irrigation, drainage, water content, aeration, temperature, nutrients, tillage practice, plant disease, and related phenomena upon root habit and their significance in crop production, the reader is referred to the same volume.

ACTIVITIES OF ROOTS IN SUBSOIL

The great extent of the root systems of most vegetable crops and their usual thorough occupancy of the subsoil may at once arouse interest concerning the importance of the deeper soil layers. Experiments have shown that the roots of crops are active in the absorption of both water and nutrients even to the maximum depth of penetration. 174,165 Nutrients, when available, are taken from the deeper soils in considerable quantities, although to a lesser extent than from the soil nearer the surface which the roots occupy first and, consequently, at least in annual crops, where they absorb for the longest time. The deeper portions of the root system are often particularly active as the crop approaches maturity. Nutrients absorbed by them may produce a pronounced effect both upon the quantity and the quality of the crop yield.

METHOD OF ROOT STUDY

In the present studies the direct method of root examination has been employed. It has been used by the writer and his coworkers in the excavation of hundred of root systems during the past 14 years and has proved very satisfactory. By the side of the plants to, be examined, a long trench is dug to a depth of about 5 feet and of convenient width. This affords an open face into which one may dig with hand pick and ice pick and thus uncover and make a careful examination of the entire root system. This apparently simple process, however, requires much practice, not a little patience, and wide experience with soil structure. In every case several plants were examined at each stage of development to insure an adequate idea of the general root habit. As the work of excavation progressed, the trench was deepened, if necessary, so that finally the soil underlying the deepest roots was removed. Frequently, the trenches reached depths of 6 to 11 feet.

Fig. 2.2 Trench used in the Study of Root Systems.

In this case 3-year-old trees were being excavated. They did not extend deeper (5 to 8 feet), however, than many vegetable crops of a single season's growth.

Upon excavating the roots, detailed notes and careful measurements were made in the field. After several plants were examined, these notes were studied and any point that remained indefinite was at once clarified by further study. This method leads to a high degree of accuracy. Drawings of the root systems were made in the field on a large drawing sheet with pencil and later retraced with India ink. They were made simultaneously with the excavation of the roots and always by exact measurements. In the drawings the roots are arranged as nearly as possible in a vertical plane; that is, each root is placed in its natural position with reference to the surface of the soil and a vertical line from the base of the plant. In some cases the drawings represent the roots in their natural position in the surface foot of soil.

In every case it was sought to illustrate the average condition of root development rather than the extreme. Although the drawings were made on a large scale, the rootlets were often so abundant that it was quite impossible to show the total number. Such drawings, however, carefully executed, represent the extent, position, and minute branching of the root system even more accurately than a photograph, for under the most favourable conditions, especially with extensive root systems, the photograph is always made at the expense of detail, many of the finer branches and root ends being obscured.

CONDITIONS FOR GROWTH

Since roots, like aboveground parts of plants, are greatly modified by environment, a brief statement will be given of the conditions under which the crops were grown.

SOIL

The vegetable crops at Lincoln were grown in a fertile, dark-brown, Carrington silt loam. Preceding them white clover had been raised for 2 years. This was followed by a crop of maize. The soil was not only in an excellent condition as regards tilth but also was well supplied with humus. In general, it may be said that the soils of eastern Nebraska have a sufficient supply of all the essential nutrients to insure good crop yields, except that, owing to the systems of cropping, there may be a deficiency in available nitrates. The area was only slightly rolling, the general slope being towards the south. As usual, some differences in soil texture were found in excavating the root systems in different parts of the area. The following description of the soil profile represents an average condition.

The surface 12 to 14 inches was a mellow, very dark silt loam from which the roots were readily removed. At greater depths it intergraded into a dark-coloured, clay subsoil. This became quite sticky when wet and hard when dry. It exhibited a columnar or jointed structure, especially below 18 inches, cracking badly when it shrank upon drying. Roots were removed with much more difficulty from this soil layer. At about 3 feet in depth the subsoil became much lighter yellow in colour and graded rather abruptly into an easily worked, quite mellow, friable soil type approaching loess in many of its physical properties. This extended beyond the depth of greatest root penetration, *i.e.,* over 12 feet. It was characterized by rusty streaks and numerous small calcareous areas and concretions. Earthworm burrows penetrated the soil to depths of 8 feet or more and numerous old-root channels extended almost as deeply. In many places extensive surface cracks reached depths of 4 to 5 feet. They were sometimes 1 inch in width. These crevices had been formed during past years of drought and were filled with rich, black, surface soil which had been washed or blown into them. Cultural operations in a large measure had also undoubtedly contributed to the filling.

NUMBER AND SIZE OF PLATS

The crops, in nearly all cases, were grown in triplicate plats so that early, midsummer, and later examinations could be made without disturbing the areas in which the plants were to make further growth. The plats, although somewhat variable in size, were in all cases large enough to permit of normal field development, each group of plants being entirely surrounded by plants of its kind. For example, the plats of lettuce, radishes, and onions were 12 feet long and 10 feet wide, the plants in the central portion of each plat being used for a single root study. In the case of the tomato, cabbage, eggplant, and other crops where each plant occupied considerable space, the plats were proportionately larger so that the plants examined were grown under the ordinary competitive conditions and mutual environment of garden crops.

TILLAGE

The field was plowed 8 inches deep early in the spring after all cornstalks, stubble, and dried weeds had been carefully removed. This was followed by repeated harrowing until an excellent, firmly compacted, moist seed bed of good soil structure was formed. On areas not immediately planted, weeds were removed by raking and hoeing. After the seeds were sown or the seedlings transplanted, further cultivation was done very shallowly with rake or hoe so as not to disturb the roots. Cultivation and weeding were repeated, as in ordinary gardening, when needed to prevent the growth of weeds and to keep the surface soil in a good condition of tilth. At no time did cultivation extend to greater depths than 1 inch.

PRECIPITATION

The rainfall of 19.2 inches during the 6 months of the growing season was 2.3 inches below the normal, with periods of moderate drought occurring in May and again in July. April with a deficiency of 1.2 inches had three well-distributed showers. In May efficient rains occurred on the eighth and fifteenth, with a monthly deficiency of 3 inches. June was a wet month with a total of 6.6 inches of precipitation, mostly in seven well-distributed rains. This was 2.3 inches above the mean. July had four rains during the latter half of the month and a total precipitation 1.8 inches below the mean. Five well-distributed showers fell in August, an excess of 1.2 over the mean. September had only 0.2 inch above normal, the rains also being well distributed.

SOIL MOISTURE

It is a well-established fact that rainfall is only a very general indicator of soil moisture, since many other factors both climatic and edaphic intervene between precipitation and water available for plant growth.

Hence, a study of the soil moisture in several of the plats was made from time to time throughout the growing season. Ideally this should have included determinations for each kind of crop at frequent intervals. Owing to the laborious task of excavating roots this was not attempted.

Table. Approximate Available Soil Moisture

		Cabbage				Beets		
Depth, Feet	**May 29**	**June 10**	**June 24**	**Aug. 4**	**May 29**	**June 10**	**June 30**	**Aug. 13**
0.0–0.5	7.6	5.4	13.1	12.5	13.0	8.8	14.9	17.5
0.5-4.0	9.0	8.0	12.5	3.4	12.5	10.2	12.6	12.8
1–2	12.3	8.9	10.4	4.0	10.0	8.0	10.4	1.6
2–3	7.5	7.2	9.3	2.9	5.6	3.8	9.7	1.0
3–4		9.6	7.7	4.0		3.9	10.0	2.1
4–5		9.0					11.6	2.6
5–6								6.7
6–7								9.0

	June 17	July 13		July 25	Aug. 22		Aug. 29	Cct. 5	
Depth, Feet	**Peas**	**Lettuce**	**Radish**	**Sweet Corn**	**Sweet Corn**	**Cucumber**	**Rutabaga**	**Rutabaga**	**Parsnip**
0.0–0.5	18.8	2.4	3.8	0.3	12.4	18.1	1.6	5.2	12.3
0.5-4.0	13.5	5.4	7.5	0.4	10.7	14.8	0.3	4.0	9.7
1–2	7.1	6.9	8.3	2.1	4.7	12.4	0.7	4.5	8.0
2–3	4.2	8.4	8.3	2.4	2.4	9.3	2.7	2.4	6.9
3–4	4.2	9.1	7.9	3.4	3.4	7.1	2.5	2.2	5.0
4–5		10.5	11.7	4.3	3.8			11.4	5.0
5–6								17.4	

An examination of Table shows that in both the cabbage and beet plats sufficient water was available until the middle of August to promote a good growth. On July 25 sweet corn, which uses water in large amounts, had rather thoroughly depleted its supply of soil moisture, especially in the surface soil, but later in the season sufficient water was again present to promote normal development. Differences in the degree of soil-moisture depletion by the various crops will be discussed when the root habits of the crops are considered.

TEMPERATURE

The temperature of the air is an important factor not only directly in promoting metabolic processes but especially in connection with root studies in modifying humidity and affecting transpiration. The amount of water lost from the aboveground parts reflects itself in the development and extent of the absorbing organs.

Table gives the average day temperatures for the several weeks. These data were obtained in the usual manner from the records of a thermograph placed in an appropriate shelter in the field. It records the shade temperature at a height of 5 inches from the soil surface. The average daily temperatures are also included. Except for a few extremely hot days, temperatures throughout were very favourable for growth.

Table. Air Temperatures in Degrees Fahrenheit During 1925

Week ending	May	May	May	June	June	June	June	July
	13	20	27	3	10	17	24	1
Average day	57.3	61.9	67.9	78.0	80.0	81.5	82.8	77.3
Average daily	49.5	57.5	60.0	74.6	74.3	77.9	78.7	72.1
Week ending	**July**	**July**	**July**	**July**	**Aug.**	**Aug.**	**Aug.**	**Aug.**
	8	15	22	29	5	12	19	25
Average day	88.4	90.1	82.0	78.2	74.0	77.4	81.2	79.4
Average daily	82.5	84.5	76.0	74.1	64.8	73.2	76.8	74.8

A continuous record of soil temperatures at a depth of 6 inches was obtained from May 16 to Aug. 25. During May the temperature ranged between 58 and 82°F., except on May 17 when a temperature of 50°F. was reached. The daily variation never exceeded 15°F. and was usually about 10°F. In June a variation from 62 to 90°F. was found, although the daily range did not usually exceed 12°F. But on one day a range of 19°F.

was recorded. The soil at this depth was warmest at 6 p.m. and coldest at 8 a.m. This condition held throughout the summer. It represented a lag of nearly 4 hours behind the maximum and minimum air temperatures, respectively. Throughout July the temperature varied between 65 and 94°F. The daily range was usually through 15°F. (maximum, 18°F.), the highest and lowest temperatures occurring as in June. In August a temperature range from 61 to 89°F. was found, the daily range being slightly greater than for July.

The soil above 6 inches was not only much warmer at. times but also subject to greater temperature fluctuations than at greater depths. Thus roots lying close to the surface of the soil were subject to the influence of an environment which was quite different from that affecting those growing more deeply.

It seems probable that surface-soil temperatures never became so high as to be injurious to root growth but that the absence of roots or their death in the shallowest soil was due to lack of a sufficient supply of moisture. 15,174 Soil temperatures are also of great importance because of their relation to disease-producing organisms that may greatly limit successful crop production.

HUMIDITY

The loss of water through transpiration is directly controlled in a large measure by the amount of moisture already in the air. This is also an important factor governing the evaporation of water directly from the surface soil. Consequently, a record of relative humidity was obtained by means of a hygrograph appropriately sheltered with the recording apparatus about 5 inches above the soil surface.

Table. Relative Humidity in Per cent, 1925

Week ending	May	May	May	June	June	June	June
	13	20	27	3	10	17	24
Average day	56.9	58.9	44.0	59.5	53.5	67.5	74.8
Average daily	65.0	67.0	55.2	65.2	62.0	75.8	81.6

Week ending	July	July	July	July	Aug.	Aug.	Aug.	Aug.
	1	15	22	29	5	12	19	25
Average day	54.7	51.8	41.7	54.2	53.1	70.2	56.7	54.8
Average daily	66.4	61.6	55.6	63.1	63.8	78.6	67.3	63.1

EVAPORATION

Evaporation was measured by Livingston's standardized, non-absorbing, white, cylindrical atmometers. These were placed in a bare area in the field with the evaporating surface at a height of 2 to 4 inches above the surface of the soil.

Table. Average Daily Evaporation

May 16-30	May 30 Jun 17	June 17-24	Jun24- Jul 11	July 11-22	July 22-29	Jul 29- Aug 5	Aug 5-12	Aug 12-25	Aug 25-29
43.7	32.0	22.0	43.5	43.7	35.5	31.5	19.3	25.4	35.0

CONDITIONS FOR GROWTH AT NORMAN, OKLA

The greater portion of root studies in Oklahoma was made during 1926. Some work was also done the preceding growing season and the root growth of several crops was followed during the intervening winter. Where root excavations were made in Oklahoma this is so stated in the text.

SOIL

The crops were grown in a fine sandy loam soil that had been heavily manured for a number of years and used for growing vegetables. It was in excellent condition as regards tilth and amount of humus.

The area was level and the soil quite uniform throughout. The surface foot had a reddish-brown colour, was quite sandy in texture but rather compact, and contained enough clay to exhibit a tendency to bake when it dried after heavy rains.

In the second foot it was a light-chocolate brown and contained more clay. When dry, it became quite hard, especially at a depth of 18 to 24 inches. The greater sand and smaller clay content of the soil prevented it from cracking and. fissuring as did the soil at Lincoln. This, as will be shown, had a marked effect upon general root habit and branching. Roots penetrated the 18- to 24-inch soil layer with some difficulty, especially when it was rather dry. This was shown by their tortuous courses.

The third foot consisted of a reddish-yellow sandy clay. It contained numerous, small, black concretions of iron 1 to 2 millimeters in diameter. After long periods of drought it became very hard. The next 18 inches were much sandier and had more and larger concretions but showed no change in colour. At greater depths the soil became still sandier, was mottled with reddish-brown spots, and contained many concretions. At 7 to 8 feet it was very compact. Table shows the sizes and proportions of the various soil particles at the several depths together with the hygroscopic coefficients.

Table. Mechanical Analyses and Hygroscopic Coefficients of Soil.

Depth of Sample, Feet	Fine Gravel Per cent	Coarse Sand, Per cent	Medium Sand Per cent	Fine Sand Per cent	Very Fine Fand Per cent	Silt Per cent	Clay Per cent	Hygroscopic Coefficient
0.0-0.5	0.00	0.66	2.28	14.01	47.14	19.84	16.07	5.2
0.5-1.0	0.00	0.65	1.70	7.54	47.14	22.50	20.46	6.7
1-2	0.00	0.55	1.97	8.66	41.56	19.59	27.67	8.5
2-3	0.00	0.41	1.35	6.55	29.45	27.96	34.26	9.7
3-4	0.00	0.61	2.42	12.05	33.78	20.96	30.16	9.2
4-5	0.00	1.07	3.66	16.33	36.67	12.96	29.30	8.9
5-6	0.00	1.08	3.94	18.31	36.30	13.39	26.97	8.5

METHOD OF PLANTING AND TILLAGE

The field was plowed to a depth of 8 inches late in March. Repeated disking and harrowing resulted in a good, compact seed bed. All of the crops were planted by hand, each in four to eight rows, 3.5 feet apart in order to permit of cultivation with a horse-drawn harrow. Shallow cultivation was given after each rain to prevent the formation of a surface crust and also to kill weed seedlings. A mulching fork, supplemented by a hoe when necessary, was used near the plants. In this manner a surface mulch 1 to 1.5 inches deep was maintained but in no case was a cultivation deeper. A good supply of moisture was present just below the mulch during almost the entire growing season.

PRECIPITATION

The rainfall of 11.5 inches during the period when the crops were grown was 3.6 inches below the mean.

April with 1.9 inches had a deficiency of 1.3 inches and no efficient rainfall after the tenth. May had 2.3 inches which was somewhat (3.1 inches) below the normal amount. No efficient rain fell between May 8 and 30. During June precipitation was 1.6 inches less than the mean. A precipitation of 2.2 inches, however, was fairly well distributed throughout the month. July had 5.1 inches rainfall, 2.4 inches above the mean, little moisture falling after the thirteenth.

SOIL MOISTURE

Notwithstanding the rather limited rainfall, a good supply of water was present throughout the entire growing season. The fine sandy loam soil of the level field, loose from the cultivation of the previous season, had been thoroughly wet by the fall and winter rains.

Water had entered the deeper subsoil and moistened it beyond the depth of greatest root penetration. Most of this water was retained in the spring and supplemented by rains in March. Thus, although the summer rainfall was light, a good supply of moisture was present throughout the entire growing season. Since soil samples were taken from the middle of the area between the rows, the earlier determinations represent the total available moisture. Later the widely spreading roots extended into these areas and the water supply was considerably reduced.

Table. Average Approximate Available Soil Moisture

Date	0.0 to 0.5 foot	0.5 to 1.0 foot	1 to 2 feet	2 to 3 feet	3 to 4 feet
Mar. 11	8.6	11.5	10.5	5.3	7.8
Mar. 18	10.6	12.2	10.1	7.5	8.3
Mar. 25	10.2	10.4	9.0	7.8	7.6
Apr. 1	14.6	15.5	12.3	9.0	7.7
Apr. 8	11.6	13.3	10.7	8.9	8.7
Apr. 15	13.9	13.9	12.2	10.3	9.5
Apr. 22	10.9	11.0	11.2	9.6	10.3
Apr. 29	9.8	12.3	10.1	9.0	10.1
May 6	13.6	14.0	11.9	10.5	9.9
May 13	14.5	14.0	11.5	7.2	10.6
May 20	10.1	12.9	10.8	9.8	9.3
May 27	15.3	10.6	10.3	9.4	10.2
June 3	9.4	11.2	10.6	9.5	9.3
June 10	3.5	7.2	5.7	7.8	7.9
June 17	6.4	13.5	10.7	10.5	10.3
June 24	8.9	11.3	9.9	9.3	9.4
July 1	7.0	6.5	9.0	7.0	8.7
July 8	12.1	7.1	8.1	6.2	7.3
July 15	8.0	4.0	8.0	5.4	7.7
July 22	4.3	3.4	5.9	5.3	8.1

An examination of Table shows that the soil was moist when the crops were planted and that sufficient water was available at all times to promote a good growth.

That conditions of soil moisture in the plats of pepper and Swiss chard were representative for the crops as a whole was shown by frequent soil sampling among the roots of other vegetables. In nearly every case very similar results were obtained.

This, however, was to be expected, since the root habits of both pepper and Swiss chard are representative and their aboveground transpiring parts moderately extensive. The excellent moisture content in the surface 6 inches was undoubtedly in part due to the soil mulch preserved at all times.

OTHER FACTORS

Another factor favourable to moisture conservation was a very cool, late spring. Warm weather did not begin until the second week in April. The surface soil warmed rapidly, there was enough moisture to promote prompt germination and vigorous growth, and the crops made excellent progress even under a light precipitation.

The usual drought and high temperatures of July were replaced by a relatively cool, moist midsummer during which the plants made a continuous growth. The average day temperatures (6 a.m. to 6 p.m., inclusive) and the average daily temperatures (24 hours), obtained by means of a thermograph placed in the field as at Lincoln, are shown in Table.

Table. Temperatures in Degrees Fahrenheit, and Average Daily Evaporation

Week Ending	March		April					May	
	18	25	1	8	15	22	29	6	13
Average day temperature	53.3	60.3	40.6	52.4	50.0	61.4	72.6	77.7	72.2
Average daily temperature	48.8	55.0	38.5	47.4	46.9	56.9	65.0	69.9	67.6
Evaporation		26.6	10.2		7.0	21.7	39.1	29.1	24.8

Week Ending	May		June					July		
	20	27	3	10	17	24	1	8	15	22
Average day temperature	84.8	90.0	83.8	86.4	85.4	80.6	86.8	94.6	86.5	86.2
Average daily temperature	76.8	87.9	78.8	79.3	79.7	75.3	79.1	89.8	82.1	81.8
Evaporation	39.1	46.4	26.1	41.7	48.0	17.3	29.0	51.4	22.7	31.0

The average daily evaporation from porous-cup atmometers, similar to those used at Lincoln, is also shown in

Table. The atmometers were placed in an uncropped area with the evaporating surface 6 to 9 inches above the ground. An examination of these data shows that during two or three periods, occurring in May, June, and July, evaporation was very high. This also indicates conditions for high transpiration.

CONDITIONS FOR GROWTH DURING WINTER

During this period many long-lived vegetable crops showed a vigorous, renewed growth. Growth was continued, especially by the hardier plants, during an unusually cool October.

Owing to favourable soil moisture and rather uniformly moderate temperatures during November and December, growth in several crops slowly continued. Following the cessation of growth in January, renewal of the development of both roots and tops was resumed during February which had a temperature of 8.2°F. above the mean.

The average day temperatures and average daily temperatures for the entire period, secured by a thermograph placed in the field, are shown in Table. Here also are included isolated readings of the soil temperature at a depth of 6 inches, as well as the precipitation. The reader may wish to refer to these environmental conditions when studying the root development of the several crops, the description of which will now be taken up.

Table. Conditions For Growth During Fall and Winter

Week Ending	August			September				October	
13	20	27	3	10	17	24	1	8	
Average day temperature, degrees Fahrenheit	79.3	84.7	79.1	81.4	85.8	60.1	81.4	76.1	67.6
Average daily temperature, degrees Fahrenheit	75.6	79.7	76.0	75.	78.7	55.1	77.5	71.8	63.4

Soil temperature, depth 6 inches, degrees Fahrenheit	82	84	84	86	84	65	79	72	71
Total precipitation, inches	0.90	0.00	0.00	0.00	1.00	5.25	1.49	1.16	1.79

Week Ending	**October**			**November**				**December**		
	15	**22**	**29**	**5**	**12**	**19**	**26**	**3**	**10**	**17**
Average day temperature, degrees Fahrenheit	58.5	58.7	48.2	50.8	53.3	56.6	55.5	54.7	48.9	51.0
Average daily temperature, degrees Fahrenheit	56.1	55.6	41.1	46.5	47.9	45.3	48.5	49.1	43.3	44.6
Soil temperature, depth 6 inches, degrees Fahrenheit.	67	62	60	61	60	62	60	57	50	47
Total precipitation, inches...	0.45	0.24	0.00	0.08	2.01	0.04	0.03	0.00	0.29	0.00

Week Ending	**December**		**January**				**February**			
	24	**31**	**7**	**14**	**21**	**28**	**4**	**11**	**18**	**25**
Average day temperature, degrees Fahrenheit	37.4	39.8	33.3	42.6	52.0	46.1	56.3	47.2	47.7	55.7
Average daily temperature, degrees Fahrenheit	32.4	33.3	29.9	38.1	47.0	41.0	48.7	40.6	41.4	47.9

Soil temperature, depth 6 inches, degrees Fahrenheit	44	38	34	36	43	43	54	48	49	52
Total precipitation, inches	0.00	0.00	0.00	0.00	0.00	0 00	0.00	0.00	0.07	0.09

3

Sweet Corn

DESCRIPTION

Sweet corn is not only one of the most common but also one of the most important of vegetable crops. Like field corn it may be grown on a wide variety of soils. A corn plant requires more space than does the individual of any of its cereal relatives. In many varieties the production of suckers or secondary stems or branches from the lower nodes is very pronounced. These branches develop their own roots.

Sweet corn of the Stowell's Evergreen variety, a standard, main-crop variety, was planted June 2. This is one of the old, well-known, and most important canning varieties. It produces a rank growth. The hills were spaced 42 inches apart and three to four plants were permitted to grow in each hill. The grains were planted 2.5 to 3 inches deep. Weeds were kept in check by repeated shallow cultivation.

EARLY DEVELOPMENT

An initial examination was made June 18, only 16 days after planting. The plants were 7 inches tall, in the sixth-leaf stage, and had a leaf spread of 8 inches. The total leaf surface averaged 45 square inches. Unlike field corn, which usually has three roots making up the primary root system, sweet corn has but one. This seed root was already 18 inches long on most of the plants examined. Thus, allowing 3 or 4 days for germination, it had grown at the rate of over 1 inch a day. The direction of growth was not downward but obliquely outward and downward. The lateral spread was sometimes more than 1 foot, but none penetrated deeper than 11 inches (Fig. 3.1). The last 3 inches of root were unbranched but the remainder was covered with branches at the rate of 25 per inch. The younger ones were scarcely 1/4 inch in length, but towards the seed they became progressively longer and many on the older portion of the root were 3 inches long. A few were even more extensive, reaching a maximum length of 10 inches. The longer and older laterals were densely rebranched with laterals seldom exceeding 1/4 inch, but the younger ones were simple. Occasionally, a second long root arose from the base of the stem near the seed. one of these is shown in the drawing.

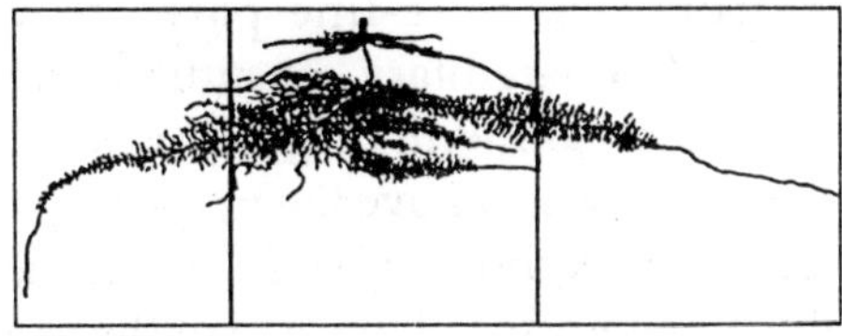

Fig. 3.1 Root System of Sweet Corn 16 Days after Planting.

In this connection it is interesting to note that recent studies on field corn have shown a positive correlation between high seminal-root production and yield. There is also some evidence that high seminal-root production tends to enhance vigour of early growth.

The secondary root system,. originating from a node on the stem about 1 inch below the soil surface, had also begun to appear. This consisted of 6 to 10 rather thick, mostly horizontally spreading roots. Apparently these were of rather recent origin but were developing rapidly. Some were 8 inches long, others just emerging from the stem. The younger and shorter ones were smooth and unbranched like the 3 to 4 inches on the ends of the older ones. Otherwise short branches occurred at the rate of 20 to 25 per inch. Root hairs were very abundant and the roots were so covered with adhering soil particles, except the shiny white ends, that they appeared very thick and black even after removal from the soil. Figure 3.1 shows that the total absorbing area of the roots really was quite large. In the case of field corn it has been found to exceed that of the tops.

Experiments with the smaller cereals, *viz.*, wheat, barley, and rye, indicate that the primary root system (seminal roots and their branches) serve largely the main stem, whereas the nodal roots chiefly serve the tillers. To what degree a similar relationship holds in sweet corn would constitute an interesting and valuable experiment.

RELATION OF ABSORBING AREA TO SOIL MOISTURE

In one experiment Nebraska White Prize dent corn of the *F*1 generation from two pure-line parental strains and, consequently, of similar hereditary constitution was grown for 5 weeks in fertile loess soil with original water contents of 9 and 19 per cent, respectively, above the hygroscopic coefficient. The plants were in the eighth-leaf stage when the roots were examined. In the more moist soil the area of the tops, including the stem and both surfaces of the leaves, was 82 per cent of that of the roots. But in the drier soil the tops had only 46 per cent as great an area as the roots. In other words, the absorbing area of the roots (exclusive of root hairs which covered the entire root system) was 1.2 times as great as the area of the tops in the more moist soil and 2.2 times as great in the drier soil.

The total length of the main roots in the two cases was about the same, as was also their diameter. In neither case did the main roots make up more than 11 per cent of the total absorbing area. In the drier soil 75 per cent of the area was furnished by the primary laterals and the remaining 14 per cent by branches from these. But in the more moist soil the primary branches furnished only 38 per cent of the root area. It seemed as though the plant had blocked out a root system quite inadequate to meet the heavy demands for absorption made by the vigorous tops, and as the soil became drier the remaining 51 per cent of the area was furnished by an excellent development of secondary and tertiary branches. Maize, in loess soil with only 2 to 3 per cent of water in excess of the hygroscopic coefficient, had, in proportion to the length of the main roots, about one-third more laterals than it had in a similar soil of medium water content. Moreover, the absorbing area in comparison to tops was greater.

Similar top-and-root area relations have been found to hold for other plants when grown in like soils of different water content. Numerous experiments have shown that root extent is greater in dry Soil. 68 Because of the difficulty of recovering the root system in its entirety from the soil and the onerous task of measuring the length and diameter of all its parts, but few data are available.

For example, a plant of maize only in the eighth-leaf stage has from 8,000 to 10,000 laterals arising from the 15 to 23 main roots. Usually the relations are stated in the ratio of the dry weight of roots to tops, an expression difficult of interpretation in terms of function. The larger, thicker, and heavier roots are least significant, the delicate branchlets, too often lost in such determinations, being of greatest importance although adding little to weight.

MIDSUMMER GROWTH

The plants had a height of about 4.5 feet and were very leafy. The parent stalk had usually given rise to two to four tillers but sometimes there were as many as six. The very numerous leaves, averaging 2.5 feet long and 3.5 inches wide at the base, offered a very large area for food manufacture and transpiration. For example, one plant with four tillers had a leaf surface of 20 square feet. The tassels were about half out.

The root system had made a really wonderful growth and was clearly in a state of very rapid development. After considerable study a plant with four tillers was selected as typically representative, and was fully examined and described. There were 45 roots of larger diameter, 3 to 4 millimeters. These were older and longer than the others. Another lot of 15 roots were about, 1.5 millimeters in diameter and also extended widely. In addition there were 33 young roots of a diameter of

1.5 to 3.5 millimeters and of an average length of 4 inches (varying from 1.2 to 8.5 inches). Finally, 29 smaller, fine roots only 0.5 millimeter thick and about 2 inches long arose from the root crown. It is impossible to represent all of these in detail in the most carefully executed drawing. But at least the principal features can be clearly portrayed.

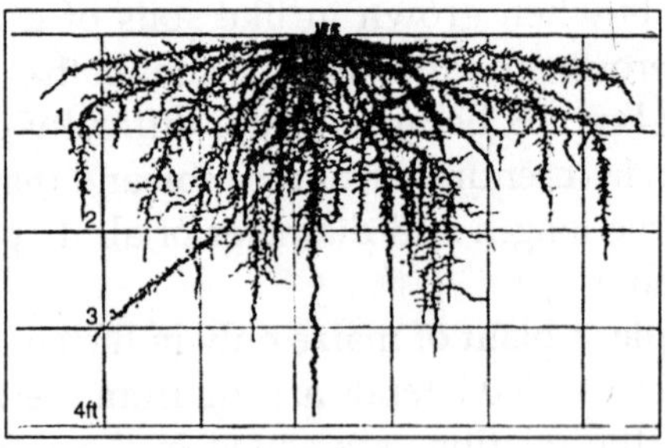

Fig. 3.2 Root System of Sweet Corn 8 Weeks Old.

The primary root (often erroneously called temporary), which could still be easily identified, pursued an obliquely downward and outward course, ending 28 inches horizontally from the base of the stalk and at a depth of 37 inches. Only a few roots penetrated deeper. One, however, was found near the 4-foot level, but the usual depth of maximum penetration scarcely exceeded 3 feet.

The working level was at a depth of 2 feet and a maximum lateral spread of 3 feet had been attained. The working level, or the working depth, means that to which many roots penetrate and at which much absorption must occur.

The thorough occupancy of even the surface 3 inches of soil and the competition of the roots between the 42-inch rows may be plainly seen. The lax, meandering course of the longer main roots of the adventitious or secondary root system was such as to ramify the soil completely from the surface to directly beneath the plant. These strong, tough roots were 3 to 4 millimeters in diameter at their origin and maintained a diameter of 1.5 to 2 millimeters to their tips. They were extremely well branched. Only the 3 to 5 inches of the rapidly growing root ends were free from branches. The branches were usually most numerous and often longest throughout the first few feet of their course. Here they were frequently distributed 18 to 25 per inch, but they were profuse throughout, *i.e.,* 8 to 12 per inch. All of the branches are not shown in the drawing. These were variable in length, the usual length of the shorter ones ranging from 0.2 to 2 inches.

But branches 6 inches to 2 to 3 feet long were also common. The shorter rootlets were often unbranched, but longer ones (mostly those over 1 inch in length) were rebranched with 20 branches, 0.1 to 1 inch long, per inch. These longer sublaterals were again rebranched. Often the larger branches on the main roots were 1 to 1.5 millimeters thick and branched like the main root. Thus, dense masses of roots and rootlets ramified the soil.

The root network was especially dense within a radius of 6 to 10 inches from the plant. Here, in addition to the older roots, new surface roots were formed. For example, one of these only 7.5 inches long was 5 millimeters thick. It tapered-to a tip 2 millimeters in diameter. There was a total of 63 branches, many of which were profusely rebranched as shown in Fig. 3.3 roots of this general type were found near the base of the plant, and some actually had 40 branches per inch! Some aerial roots, as yet without branches, extended 2 to 4 inches into the rootfilled soil. All of the surface roots were very turgid and brittle. In fact, the denseness and abundance of the rapidly growing roots can scarcely be overemphasized.

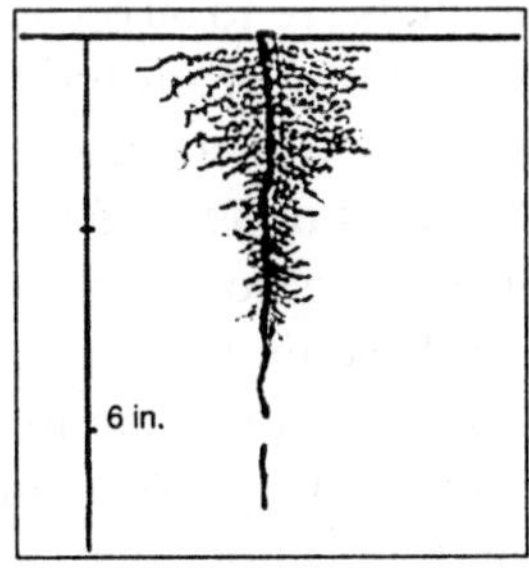

Fig. 3.3 One of the Younger and Shorter but Densely Branched Roots of Corn.

MATURING PLANTS

Late in August a final examination was made. The plants had an average height of 5.3 feet; some stalks were 6.5 feet tall. The leaves had not increased in average size but there were more of them owing to the development of the tillers, some of which were now as large as the parent plant. Only a few of the basal leaves had deteriorated and the plants were in good condition. The one selected for detailed study had five stalks and seven ears. The kernels were well filled but the husks were only beginning to dry.

The root system had developed proportionately to the tops. Branchlets extended to the root tips and indicated that growth was practically complete. The surface area delimited at the former examination (approximately 3 feet on all sides of the plant) had not been greatly extended, but the soil volume, of which this was one end, had been greatly extended in depth. A working level of 50 inches was found and numerous roots ended between 60 and 68 inches, which was the maximum depth.

After continued study a typical plant was selected and its root system worked out in the usual manner. A total of 78 main roots (*i.e.*, roots of large diameter, 3.5 millimeters or more) arose from the base of this plant and 123 smaller ones. Those of the latter group ranged from 0.5 to 2 millimeters thick. Nine of the large main roots extended outward 18 to 42 inches from the base of the plant in the surface soil and then, turning downward, reached depths of 3 to 5 feet.

Twenty-five main roots (20 to 30 on other plants) extended outward only a short distance from the base of the plant or, more usually, ran obliquely downward in such a manner that even at their ends (at depths of 3.5 to 5 feet) they were only 12 to 18 inches from a vertical line from the base of the plant. Figure 3.4 shows that some of these more vertically descending roots were half grown at the earlier examination. But many more of them had developed at this time. The remainder of the large roots consisted of prop roots and others with a large diameter (not infrequently 5 millimeters) which mostly extended only a short distance, usually not over 12 inches, from the base of the plant. As regards the remaining very numerous but smaller roots, these extended outward almost entirely in the surface foot of soil. Here they ended at distances of 3 to 24 inches from the base of the plant. Thus, the surface soil, especially the first 10 inches and within a radius of 9 inches from the crown, was filled with a dense tangle of roots, a network so complex and so profuse that there seemed actually to be more roots than soil. Certainly no moisture could evaporate from this soil volume. Indeed it seems many more roots were present than were necessary to exhaust it of its water and readily available nutrients! Figure 3.4 shows a surface view of this root complex, the soil having been removed and the roots exposed to a depth of 12 inches.

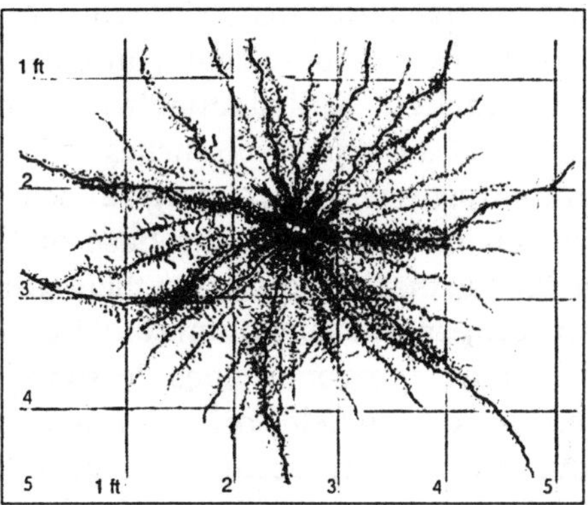

Fig. 3.4 Roots of a Mature Plant of Sweet Corn Found in the 12 inches of Surface Soil.

As regards degree of branching, it was indeed profuse. On the main, long, widely spreading roots, laterals occurred as before at the rate of 8 to 12 branches per inch, but 30 branches per inch were not uncommon. Even at depths of 4 feet or more 5 to 10 short branches per inch were usual. The branches, moreover, were, in general, longer than at the earlier examination; in fact long branches (over 5 inches) were much more numerous, and the network of branches due to rebranching much more profuse. Similar branching occurred on the long, rather vertically descending main roots, but these, like the others, were less profusely branched below 3 feet. From the tough, yellowish, cord-like roots many long branches filled the soil. Some of these in the deeper soil were quite yellow in colour. Branching of the finer surface laterals occurred at the rate indicated in the drawing. Here also the length of the branches and their degree of rebranching are shown in detail.

If one can visualize the root system of the sweet-corn plant, he will see the surface 12 inches of soil near the plant filled with a tangled root mass even to within 1 inch of the soil surface. Extending beyond this to distances of 18 to 42 inches, main horizontal roots pursue their course. From these long branches extend into the deeper soil and finally the main roots themselves turn downward and penetrate deeply. Below the 10-inch level, roots are still abundant, but not so dense, until the working level at 50 inches is reached. Still deeper roots are fewer but occasionally occur at a depth of 68 inches. The main vertical roots penetrating downward, spreading 12 inches or more on all sides of a vertical line from the base of the plant, go thoroughly fill the soil that in comparison there seems to be more or less of a gap between this soil volume and that so completely occupied by the main horizontal ones. In some plants this is less pronounced since, as in field corn, the roots spread at wider angles and thus thoroughly occupy this portion of the soil volume.

SUMMARY

Sweet corn is an important vegetable crop. The primary root system, unlike that of field corn, usually consists of a single, much-branched root. During the first few weeks this grows outward and downward over 1 inch a day. The secondary root system consists of numerous horizontally spreading roots which arise from the nodes of the stem just beneath the soil surface. The roots are so profusely furnished with laterals that the absorbing area; exceeds the area of the tops. In rather dry soil the absorbing surface is much greater than in soil that is more moist. Plants only in the eighth-leaf stage have from 7,000 to

10,000 branches. At the time of tillering both the primary and secondary root systems are still vigorously growing. The root system is composed of about 100 thick, coarse roots, about a third of which have just commenced growth, and approximately one-third as many finer ones. They spread 3 feet on all sides of the plant and thoroughly occupy the soil, interlacing between the rows, and reach a working depth of 2 feet. Maturing plants show 6 inches greater lateral spread and the soil is ramified to a depth of 5 feet.

Thus, over 180 cubic feet of soil are occupied by the roots of a single plant. Branching is profuse, 8 to 30 long, much rebranched laterals occurring per inch of main root. Briefly, sweet-corn roots extend laterally more than half as far as the stalk extends upward, and the root depth is equal to the height of this rather imposing vegetable crop.

COMPARISON WITH ROOTS OF FIELD CORN

The root system of sweet corn is very similar to that of field corn. Field corn is usually of larger stature, has a greater lateral spread (often 3.5 feet on all sides of the plant), and penetrates more deeply, usually to 5 and sometimes to 8 feet. This deeply rooting habit has been observed in Nebraska, Colorado, Kansas, Wisconsin, Illinois, and New York. That it is greatly modified by irrigation, drainage fertilizers, etc. has also been ascertained. The root system of sweet corn is so similar to that of field corn and the plants are so closely related that it is believed that similar modifications of the soil environment would induce like responses in root development.Further investigations will undoubtedly show that different varieties exhibit differences in rooting habit.

Such information is of great scientific and practical value. For example, it has been shown that inbred strains of field corn differ greatly in the character and extent of their root systems. "Certain strains... have such a limited and inefficient root system that they are unable to function normally during the hot days of July and August, when the soil moisture is low." Different strains show differences in resistance and susceptibility to root rot. It has further been shown that, in general, weak-rooted strains when compared with better-rooted ones, are more likely to lodge and give a lower yield of grain.

RELATION OF ROOT HABITS TO CROP PRODUCTION

A study of the root habit clearly shows why corn does best on a deep, well-drained soil which has an abundant and uniform supply of water throughout the growing season.

Cultivation

If the soil is well prepared before planting, the main benefits of cultivation are derived from keeping down weeds, preventing the crusting of the surface, and keeping the soil receptive to rainfall. The superficial position of the roots clearly shows why deep cultivation is harmful. Fortunately, weeds are most easily destroyed when coming through the surface of the soil by shallow cultivation such as harrowing surface-planted corn. This also breaks the soil crust, gives a drier and warmer soil, and a more vigorous crop results. Thorough preparation of the seed bed and shallow cultivation make a good combination. The harmful effect of letting weeds grow for a time is not entirely due to rapid removal of water and plant-food material by them from the soil but to the breaking of the corn roots in the deeper cultivation necessary for weed eradication.

On heavy soils, however, a slight benefit from cultivation other than weed control may be gained by better aeration. As the oxygen supply in the soil air is decreased, rate of growth diminishes in a soil with a high temperature. For example, corn roots, in a soil atmosphere of 96.4 per cent nitrogen and only 3.6 per cent oxygen, at a temperature of 30°C. grow about one-third as rapidly as at the same temperature under normal conditions of aeration. But at 18°C., growth is increased to about two-thirds the normal rate at that temperature when the soil is well aerated.

An examination of the half-grown root system explains why late tillage, except for weed eradication, is of little value. The roots are so well distributed through the soil that little moisture will escape even from uncultivated land. Hilling at the last cultivation not only acts as a mechanical support to the stem but also encourages the development of brace roots which are an additional aid in holding the plant up against strong winds. But, if hilled early, later cultivation partly removes the hill and exposes a portion of the root system. Even shallow cultivation cuts many of the roots and deep cultivation is very harmful and greatly decreases the yield. Cultivation of field corn to a depth of 4 inches during a period of 9 years in Ohio gave a decreased yield every season but one, as compared with similar cultivation to a depth of 1.5 inches. The average decrease per acre was 4 bushels of grain and 183 pounds of fodder. In Indiana, very similar results have been obtained. In Missouri, deep cultivation compared with shallow reduced the yield 6.5 to 13 bushels per acre. The harmful effects of deep cultivation are always more pronounced during years of drought. When the plants begin to shade the ground, wind movement is reduced, evaporation is decreased, and the thick network of roots near the surface absorbs the water and prevents its escape into the air.

As a result of a series of cooperative experiments carried on in 28 states during a period of 6 years, it has been found that as large yields of grain were gotten by keeping down the weeds by cutting them at the surface of the soil without forming a mulch as by cultivation. Further analysis of these findings showed that in subhumid or semiarid sections the average yield from the uncultivated land was only 85.9 per cent of that from the cultivated. Thus, the greater need for cultivation in such sections, compared to humid ones, is illustrated. The yields without cultivation, moreover, were found to be relatively higher on sandy loams and silt loams than on clays or clay loams. A difference of 13 per cent was determined between the sandy loams and clays. 109 For the highest yields, cultivation should never be deep enough to injure the roots seriously.

FERTILIZERS

Hill fertilizing of corn promotes more vigorous early

vegetative development with earlier tasseling and earing. The observation of farmers that corn fertilized in the hill sometimes suffers more from drought than when grown in soil where the fertilizer has been uniformly distributed may be explained by a study of root extent in relation to tops. Although no differences were found in the actual abundance, depth, or lateral spread of the roots, the more luxuriant plants resulting from hill fertilizing had a relatively smaller root system. This may also explain why, in Missouri, applying fertilizer in the hill or row yields good returns during seasons of abundant rainfall, but in dry seasons there is more danger that the fertilizer may cause the corn to "fire" than when it is applied ahead of the planter with a fertilizer drill. Because of the extensive development of the roots of practically all cultivated plants, it seems probable that the chief effect of hill manuring is to promote vigorous early growth and that the plant receives little benefit from the manure at the time it is completing its growth and maturing its seed.

Experiments have shown that corn absorbs nitrates and undoubtedly other nutrients at all depths to which the roots penetrate. Competition for water and nutrients from the interlacing roots of plants in the same hill or in adjacent rows is often very severe. These root relations should be considered in the rate of planting especially on less favourable soils and in dry climates.

4

Onion

INTRODUCTION

The common onion (*Allium cepa*) is a biennial with large bulbs that are usually single. It is the most important bulb crop and is exceeded in value by only four other vegetable crops grown in the United States, *viz.*, potatoes, sweet potatoes, tomatoes, and cabbages.

Like all the other bulb crops it is hardy and, in the North, is planted very early in the spring. In the South it is grown as a winter crop. The bulk of the crop is grown from seed sown in the open, but the plants are also grown from transplanted seedlings that have been started under glass or in a seed bed and by planting sets.

SOUTHPORT WHITE GLOBE ONION

For root examination the Southport White Globe onion was planted Apr. 10. Especial care was exercised to have the soil fine and loose and with a smooth surface, since the seeds are small and do not germinate quickly. Thorough preparation of the soil is an essential feature in the successful growth of nearly all crops. It is especially important in the production of vegetables. The seed was sown in drill rows in the field where the crop was to mature. The rows were 14 inches apart and the seed was covered 1/2 inch deep. From time to time the soil was shallowly hoed and all weeds were removed from the plots.

EARLY DEVELOPMENT

The onion develops a primary root which, under very favourable conditions, may reach a length of 3 to 4 inches 10 days after the seed is planted. In the meantime the cotyledon comes from the ground in the form of a closed loop. By the time the first foliage leaf emerges from the base of the cotyledon, several new roots make their appearance near the base of the stem. The first field examination was made June 4. The plants were about a foot tall and had an average of four leaves each. These varied from 4 to 12 inches in length and 2 to 5 millimeters in diameter. The total area of the cylindrical leaves per plant averaged 8.5 square inches. Each plant was furnished with 10 to 12 delicate, shining white, rather poorly branched roots (Fig. 4.1). The longest usually pursued a rather vertically downward course to a maximum depth of 12 inches. The lateral spread from the base of the bulbous stern did not exceed 4 inches. The roots were rather poorly branched and frequently somewhat curled or pursued a zigzag course.

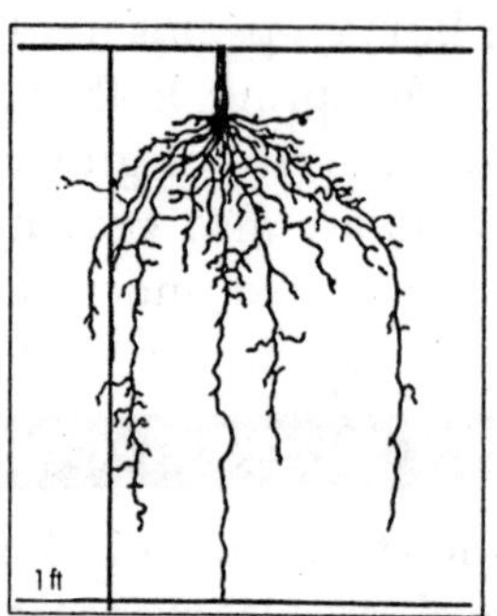

Fig. 4.1 An Onion 8 Weeks after the Seed was Planted.

EFFECT OF SOIL STRUCTURE ON ROOT DEVELOPMENT

To determine the effect of loose and compact soil on root growth, onions were grown in rectangular containers with a capacity of 2 cubic feet and a cross-sectional area of 1 square foot. A fertile, sandy loam soil of optimum water content was screened and thoroughly mixed and thus well aerated. One container was filled with very little compacting of the soil. It held 173 pounds. Into the second container 232 pounds of the soil were firmly compacted.

Surface evaporation was reduced by means of a thin, sand mulch. Onion seeds were planted in each container early in April, and a month later the sides were cut away from the containers and the root systems examined. Owing to clear weather and favourable greenhouse temperatures the plants had grown rapidly and were in the third-leaf stage.

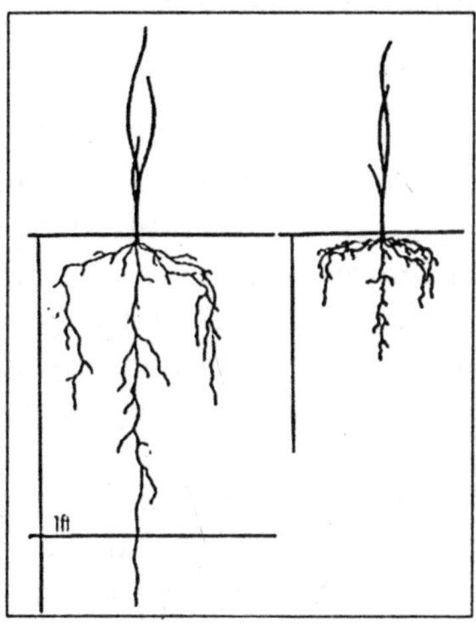

Fig. 4.2 Onion Seedlings of the Same Age. The One on the Right was Grown in Compact Soil and that on the Left in Loose Soil. Both Drawings are made to the same Scale.

The effects on root growth are shown in Fig.4.2. In the hard soil the plants nearly always possessed six roots but only five were found on those in the loose soil. Under the former conditions the vertically descending roots reached depths not exceeding 5 inches; the others spread laterally 1 to 1.5 inches and then, turning downward, penetrated to a total depth of 2.5 to 3 inches. In the loose soil one root from each plant grew downward to a depth of 12 to 15 inches; the others spread laterally 1 to 2 inches and then turned downward and reached a maximum depth of 7 inches. Thus the root system in the loose soil was not only deeper but more widely spread. The roots in the hard soil, moreover, were much more kinky. In the loose soil they made long, gradual curves; in the compact, hard soil, short abrupt ones.

Under the latter soil condition the root ends were often thickened. The roots running laterally from the plant were more horizontal in the hard soil during the first 1 to 2 inches of their course. Thus this portion of the root system was shallower than in loose soil. Branches were shorter throughout. Similar results were obtained with lettuce seedlings.

HALF-GROWN PLANTS

The longest frequently maintained their initial diameter for distances of 5 to 10 inches; others quickly tapered to a thickness of only ½ millimeter. In fact the roots showed considerable variability in this character, sometimes tapering only to enlarge again. The deepest roots penetrated vertically downward or ran obliquely outward for only a few inches and then turned downward. A working level of 20 inches and a maximum depth of 27 inches were attained.

Some of the main roots ran outward, almost parallel to the soil surface, to distances of 6 to 8 inches before turning downward at an angle of about 45 degrees. These had a maximum lateral spread of about 12 inches on all sides of the plant. Between these horizontal roots and the vertically descending ones, the soil volume thus delimited was filled with numerous roots Which extended outward to various distances and then turned downward, or pursued an outward and downward course throughout their entire extent.

As a whole the main roots were poorly branched. In the surface 8 to 10 inches of soil branching occurred at the rate of 2 to 6 laterals per inch of main root, although sometimes as many as 12 were found. Many of the laterals were only 0.5 to 1 inch long, but some reached lengths of 4 to 5 inches.

Below 8 inches depth, branches were often fewer and usually shorter, seldom exceeding 1 inch in length. The last few inches of the rapidly growing roots were unbranched. All of the laterals were slender, white, and entirely destitute of branches.

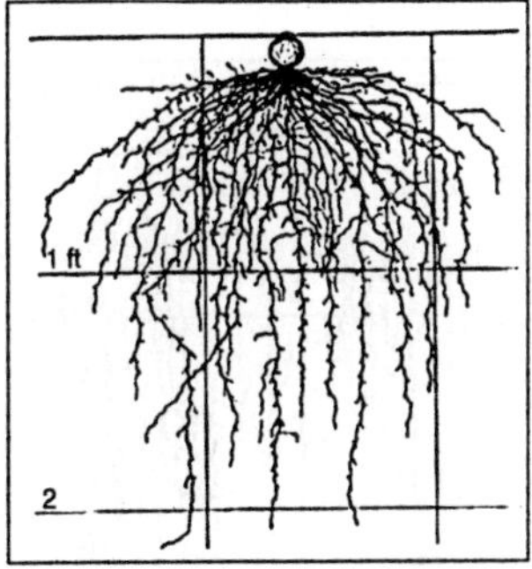

Fig. 4.3 Fibrous Root System of Onion, 3.5 Months Old.

Usually they were much kinked and curved and as often extended outward or upward as downward. The absence of roots in the surface 1 or 2 inches of soil is an important character in relation to cultivation. In this respect the onion is quite different from many garden crops.

MATURE PLANTS

From 20 to 25 roots arose from the base of the bulb. A few ran vertically downward but most of them ran outward at various angles, even to near the horizontal, and then gradually turned downward. The volume of soil delimited at the previous excavation (which had an area on the surface of about 4 square feet) had not been increased except in depth. The former working level of 20 inches had been extended to 32 inches. A maximum depth of 39 inches was found.

The uncompleted root growth was shown by the 2 to 5 inches of unbranched or poorly branched, glistening white, turgid root ends. The main roots were very uniform in appearance and about 2 millimeters in diameter throughout their course. They were somewhat kinked and curved, perhaps owing in part to the difficulty in penetrating the rather compact soil. The branching was somewhat uniform throughout and at the rate of three to five, rarely more, branches per inch. These were usually only 0.5 to 1 inch in length although sometimes they reached lengths of 3 to 5 inches. No secondary branches were found.

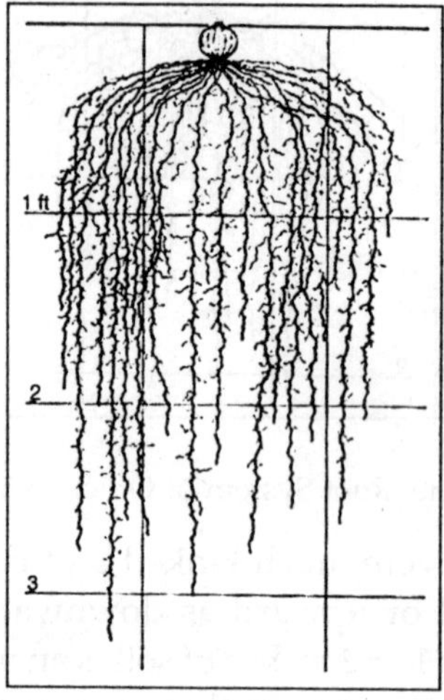

Fig. 4.4 A Maturing Onion Excavated.

The laterals, as before, were much kinked and curved. Many of them ran in a generally horizontal direction; others extended upward throughout their entire length; still others turned downward or outward and then downward. The surface 1 or 2 inches of soil were entirely free from roots. Compared with most garden crops the onion has a rather meager root system not only in regard to lateral spread and depth but also in degree of branching.

DEATH OF THE OLDER ROOTS

The decrease in the number of roots from 28 to 33 on July 25 to 20 to 25 on Aug. 21 is of interest for it is one of the few cases found among vegetable crops where the roots die before the plant approaches or reaches maturity. That death of the older roots originating from the center of the bulb actually occurs was further confirmed by greenhouse studies.

Plants were grown from seed in appropriate containers which held 2 cubic feet of soil. When the plants were nearly 3 months old and some roots had attained a depth of 22 inches, the sides of the containers were cut away and the roots examined. From one to three roots per plant arising from the center of the base of the bulb were dead. These were surrounded by six to eight living roots. The dead roots, especially one of them, held a more vertically downward course than the live ones. Similar observations have been made in both water and soil cultures. "It was observed that the roots formed at the time of germination died at about the time of the formation of the bulb, and that new ones then developed and carried the plants through their complete life cycle." Almost all of the original roots of plants which germinated May 12 were dead 2 months later (2.5 months where grown in soil), at which time the new roots were 5 inches long.

YELLOW BERMUDA ONION

Onions of the Yellow Bermuda variety were studied at Norman, Okla. The seeds were sown Mar. 16, ½ inch deep, and rather thickly to insure a good stand. When the plants were well established they were thinned to 4 inches distant in the row.

EARLY DEVELOPMENT

Although growth had been slow yet by May 9, when the first root excavations were made, the plants were 8 inches tall and had four vigorous leaves. The root system consisted of 16 to 22 delicate, white, fibrous roots about 0.5 millimeter in diameter. They varied in direction of growth from horizontal to vertical. The roots were rather lax and their course in the soil wavy. The maximum lateral spread was 10 inches and the maximum depth of penetration 12 inches. All except the shortest roots were clothed with unbranched laterals which reached a maximum length of 1 inch, becoming shorter near the root ends. These were rather evenly distributed at the rate of 2 to 4 per inch. The last 2 to 5 inches of rapidly growing main roots were free from branches.

HALF-GROWN PLANTS

A second examination was made June 14 when the plants were 16 inches tall. The bulbs were 1.5 inches in diameter and the plants examined had five large leaves. The number of roots had increased to 32, the lateral spread to 16 inches, and the maximum depth to 22 inches. The roots were also much larger than before, with an average diameter of 1 millimeter which they maintained throughout their course. Many of the formerly horizontal or obliquely descending roots had now turned downward at various distances from the bulbs. Branching was at about the same rate as before but much more irregular. The longest branches reached 2 inches. None of the laterals examined were rebranched. On some plants the laterals occurred to within 0.5 inch of the bulb; on others an unbranched area of 2 to 3 inches was found. The cause for this difference was not ascertained.

GROWTH DURING THE WINTER AND SECOND SPRING AND SUMMER

By the middle of December, the plants had developed new tops about 3 inches tall. These consisted usually of three shoots bearing three leaves each. No relation was found between the size of the bulbs and the number of roots to which they had given rise. Bulbs 2.5 inches in diameter bore from 28 to 97 roots. On a plant with 53 roots their lengths were as follows: 9 were 1 inch or less long, 14 ranged between 1 and 10 inches, and the remainder between 11 and 21 inches in length. They were 1 to 1.5 millimeters in diameter. Their distribution from horizontal to nearly vertical was identical with the root systems already described. The distribution of lateral roots was very irregular. They occurred only on the longer roots and within 1 foot of the bulb. The longest were 2 inches. All of the roots were white and rapidly growing.

Further examination was made on Mar. 9, when the leaves were 6 to 9 inches tall. The number of roots varied from 64 to 207. The soil was well filled with roots within a radius of 18 inches from the plants and to a depth of 18 inches. A maximum lateral spread of 22 inches and a maximum depth of 33 inches were found. Many of these fibrous roots pursued an outward course within 3 inches of the soil surface. At distances of 16 to 20 inches from their origin the larger, horizontal roots usually, but not always, turned downward. Lateral roots, for the most part, were more or less at right angles to the main ones, but considerable variation occurred. They formed a rather complete network of glistening white rootlets to within ½ inch of the soil surface. Even at a distance of 16 inches from the bulb many were 2 to 4 inches in length but quite unbranched. Nearer the base of the plant laterals 8 to 10 inches long were not infrequent. All of the primary laterals, except a few of the longest, were devoid of branches.

By June 20 the plant had an average height of 40 inches, although some exceeded this by 1 foot. Since each bud on the original bulb had developed into an individual plant, a hill now consisted of two to five plants with bulbs 2 inches in diameter but still loosely united by the crowns. In part the heads consisted of many flowers in full bloom but mostly of nearly mature seed pods.

Each stalk was furnished with 32 to 48 roots, the strongest and most profusely branched of any yet observed. Many were 3 millimeters in diameter, often retaining a thickness of 2 millimeters throughout their course. The roots, however, had extended their area of occupation but little to a maximum lateral distance of 22 inches and to a depth of 34 inches. This lack of linear growth had been met in two ways. New roots had replaced some of the older ones which had died and the branches throughout were longer and more profuse than heretofore. In fact the whole root system from near the bulb to within a few inches of the root tips was well supplied with branches.

Many of these were I millimeter in diameter and frequently 10 to 12 inches in length. Occasionally, they bore sublaterals. These were never abundant. On a few roots they were found to occur at the rate of two per inch throughout a distance of 3 or 4 inches. They were about 0.5 millimeter in diameter and averaged about ½ inch in length.

SUMMARY

The onion has a fibrous root system consisting of 20 to 200 shining white, rather thick roots. Some of these spread horizontally just beneath the surface soil 12 to 18 inches on all sides of the plant before turning downward. The soil area to be occupied is blocked out by the plant rather early in its development, and later growth consists chiefly in root elongation.

In compact soil the roots are shorter, do not spread so widely, and have shorter branches than in loose soil. The rate of root growth of the Southport White Globe onion is shown in Table. Depths of 18 to 32 inches are usual. The roots are poorly furnished with rather short laterals which are usually unbranched As, the older, centrally placed roots die, they are replaced by peripheral new ones. Compared with most garden crops, the onion has a rather meager root system not only in regard to lateral spread and depth but also in degree of branching.

Table. Rate of Root Growth of Onion

Age, Days	Lateral Spread, Inches	Working Depth, Inches	Maximum Depth, Inches
55	4	10	12
106	12	20	27
133	12	32	39

OTHER INVESTIGATIONS ON ONIONS

At Geneva, N. Y., onions of the Blood Red variety were examined in the middle of September. The plants were grown in a fertile, clay loam soil overlying a tenacious, gravelly clay subsoil. It was found that... the root system is by no means extensive but it is very much concentrated. The roots seem to take complete possession of the soil beneath a, circle about 8 inches in diameter to a depth of about 10 inches. The roots of the Large Red onion were washed from the soil at the same station on June 21. The leaves were but 8 inches long and the bulb only a few millimeters thick, but the roots were found to extend to a depth of 16 inches. Further examination, in September, showed that most of the roots extended no deeper than at the June examination. In a few cases roots were found at a depth of 18 inches but no horizontal roots could be traced farther than 1 foot.

The roots radiated from the base of the bulb in all directions below horizontal, some lying no more than 1 inch beneath the surface. They were of equal size throughout their length, except that they were slightly swollen close to the terminus. New roots appeared to be growing, as some were decidedly shorter than the others... The branches were all short and never subdivided.

In order to note whether the root growth in this plant the second year differs from that of the first, the roots of a plant of which the full-grown bulb was planted out in the spring were washed out June 10. The system was the same, in general, as that of the first year. An estimate made it probable that the plant had formed at least 400 feet of roots and root branches, though it had been set out but about 40 days.

Very few roots of onions grown on gravelly sandy loam soil at Ithaca, N. Y., reached a depth of more than 10 inches. A few were found as deep as 20 inches. The greatest lateral spread was 12 inches but few reached out more than 6 inches. The main root zone was found within a radius of 6 inches.

The growth of seedling Yellow Globe onions at the same station was as follows: "At the time when the onion seedlings were set out, each plant had about ten to twelve roots averaging about 3 inches long. Twenty-five days after the plants were set in the field, the roots had reached a depth of only about 2 inches and had a lateral spread of 3 inches. At this time there was an outer whorl of about ten to fourteen coarse roots without branches. Inside of this whorl were about twelve to fourteen finer roots with a few short branches. These smaller roots grew downward. Twenty days later, when the plants were 12 inches tall and a little larger at the base than a lead pencil, the roots had reached a depth of 4 inches. At this time there were about ten to twelve roots growing vertically downward and about twenty-two to twenty-four larger lateral roots extending from 3 to 4 inches from the plant. Ten days later a few roots had reached a depth of 10 inches and a lateral spread of 6 inches, although most of the growth was within from 3 to 4 inches of the plant. At this time there were about twenty-eight to thirty coarse roots with a very few short branches, and twelve finer roots, each having several branches, growing downward from the base of the plant. When the bulbs were 2 inches in diameter the greatest depth reached was 18 inches and few had gone lower than 12 inches.

The greatest lateral spread was 9 inches, with very few roots extending to more than 6 inches from the plant. At this time there were thirty-six coarse roots at the base of the outer scales and twelve finer-branched ones growing out of the center of the bulb at the base. When the bulb was full-grown a few roots had reached a depth of 20 inches, but most of them were in the surface 6 to 8 inches of soil and within a radius of 6 inches: Many roots were found very near the surface, especially close to the plant, but the greatest number were found at a depth of from 2 to 3 inches. The tendency of growth of the outer whorl of roots is outward and downward for a few inches, and then nearly vertically downward.

RELATION OF ROOT SYSTEM TO CULTURAL PRACTICE

The relatively meager root system and correspondingly limited extent of above-ground parts explain why onions can endure crowding better than most vegetable crops. Plants thrive when grown only 3 to 4 inches apart in rows 12 to 15 inches distant. In fact close planting increases the yield. The concentration of the roots into a relatively small area helps to make clear the marked response to fertilizers worked into the surface soil and also why a very fertile soil is necessary for abundant yields.

Few crops require such thorough seed-bed preparation as does the onion and this is undoubtedly due in part to the poor rooting habit of the seedlings. The roots have difficulty in penetrating stiff clay soils and this is one reason why such soils are unsatisfactory unless there is sufficient humus present to lighten them. Neither are the plants adapted to a light, open, gravelly soil, as the roots must grow in a continually moist soil, a good supply of water being especially necessary during early stages of growth. A soil retentive enough to keep a constant supply of moisture about the roots but friable enough to permit expansion of the bulbs and easy root penetration is ideal.

The present-day tendency to sow the seeds more thinly and dispense with thinning is certainly a logical practice as regards the lack of disturbance of a poorly branched root system. Pulling out plants in the process of thinning undoubtedly disturbs the root system and breaks many of the roots of those left in the rows.

Thinning when the soil is moist causes less harm. In fact one of the most harmful effects of weeds may be not so much that of shading or absorbing nutrients and water which should be used by the crop as that of root disturbance of the cultivated crop occasioned by their removal. If the root system is thus much disturbed, the onions are likely to ripen prematurely before reaching normal size.

Weeds should be destroyed as soon as they appear above the soil surface, for then their roots are poorly established. The lack of the dense network of laterals, associated with the roots of most vegetable crops, probably accounts for the need of an especially well-prepared, deep seed bed and the inability of onions to compete successfully with weeds.

Actual practice has shown that to produce a good yield of onions the weeds must be kept under control and a surface mulch be maintained to conserve moisture. For many densely rooted crops, experiments in New York have shown that cultivation for the purpose of moisture conservation is not profitable.

Corn, for example, has such an extensively developed network of roots in the surface as well as in the deeper soil that the roots absorb the moisture before it reaches the surface.

Likewise, carrots, beets, cabbage, and beans, all deeply and densely rooted.plants, show very little or no advantage in cultivation, as regards moisture content of the soil, over those in plats where the weeds were removed by scraping the surface soil. Although a difference of only about 1 per cent more moisture was found in the plats cultivated once a week, the yield of onions was greater than in plats where the soil was scraped to keep down the weeds.

In these experiments, on a gravelly sandy loam, the onions were grown in rows 18 inches apart and 4 inches distant in the row. Since, in this experiment, the main root zone was within 6 inches of the plant, a space 6 to 12 inches wide between the rows contained very few roots. Moreover, since the tops shade the soil but little, it may be clearly seen how a soil mulch resulting from cultivation would conserve the soil moisture. The roots are not so superficially placed as to be injured by shallow cultivation.

There is apparently a correlation between the extent and distribution of the root system and the response of the crops to cultivation for purposes other than the destruction of weeds, moisture being conserved by cultivation for plants like the onion with meager root development. Thus it would seem that the usual distance between the rows, about 14 inches for hand cultivation and 18 to 24 inches for cultivation with a horse-drawn cultivator, is quite ample for full root development with little or no competition between plants in adjacent rows. In fact the bulb crops are about the only vegetable crops for which this statement is true.

When onions are transplanted in the "new onion culture," it is usual to trim both tops and roots to a considerable extent. Trimming-the tops reduces the transpiring area and makes reestablishment more certain. The roots are trimmed to facilitate planting and to avoid having any long roots curl upward.

In growing onion sets, the older practice was to sow the seed late in the season on poor soil. Lack of sufficient water and nutrients, resulting in part from thick planting, combined with hot weather, resulted in the desired small bulbs. Now many commercial growers depend upon competition alone. Seeds are sown very thickly on rich soil, often 200 per foot of drill. The rows are spaced only about 1 foot apart. Under these conditions, it is impossible for the competing roots to make a normal growth and secure enough water and nutrients to supply the top which is dwarfed as a result.

5

Leek, Garlic and Asparagus

LEEK

Leek (*Allium porrum*) is a very robust, biennial plant. Bulbs are nearly always absent, the plant being grown for its blanched stems and leaves. The plants are propagated entirely from seed which is sometimes sown in the greenhouse or hotbed and the seedlings later transplanted into the field.

EARLY DEVELOPMENT

The seeds were thickly sown to insure a stand, and later the plantlets were thinned to 4 inches distant in the rows which were 3.5 feet apart. Weather conditions were not favourable for germination or growth, and the plants developed slowly. When they were nearly 2 months old and had reached a height of 10 inches, the first root examination was made. Each plant had four leaves approximately 1/2 inch in width. An average of 16 roots, in all stages of development, was determined per plant. As a rule, three were less than 1 inch long; four were 6 inches or less in length; the remainder grew to lengths of 7 to 14 inches. Their direction of growth was very similar to that of the onion, although, in general, they spread to a greater extent (Fig. 5.1).

A maximum lateral spread of 8 inches and a depth of 14 inches were found. The white, tender roots were free from abrupt turns and short curves but pursued a rather wavy course through the mellow soil. They were 1 millimeter thick and the older ones had given rise to short laterals at the rate of about two per inch.

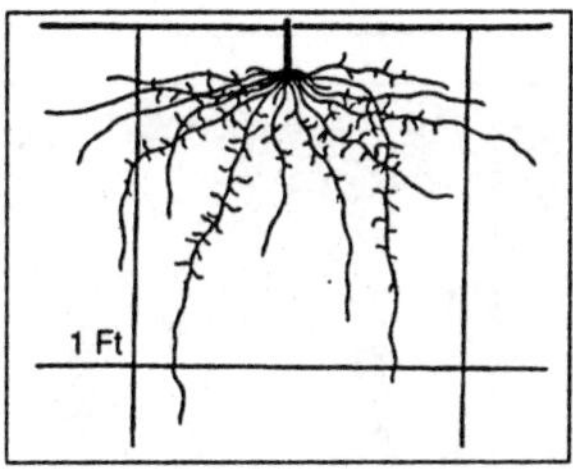

Fig. 5.1 Roots of a Two-months-old Plant of Leek.

HALF-GROWN PLANTS

From this time the plants grew rapidly, owing to a favourable supply of soil moisture. By the middle of June they were 15 inches tall, possessed bulbs 0.5 to 0.8 inch thick and leaves 1 inch in width. The roots had increased in number to an average of 38 per plant. Most of them ran 6 to 8 inches more or less horizontally outward and then downward; the others ran obliquely or, frequently, nearly vertically downward (Fig, 5.2). A maximum spread of 14 inches was attained at a depth of 6 to 8 inches and some of the vertical roots reached a depth of 20 inches. In penetrating the more compact subsoil, the roots often pursued a tortuous course and were more kinked than in the first 8 inches of mellow soil.

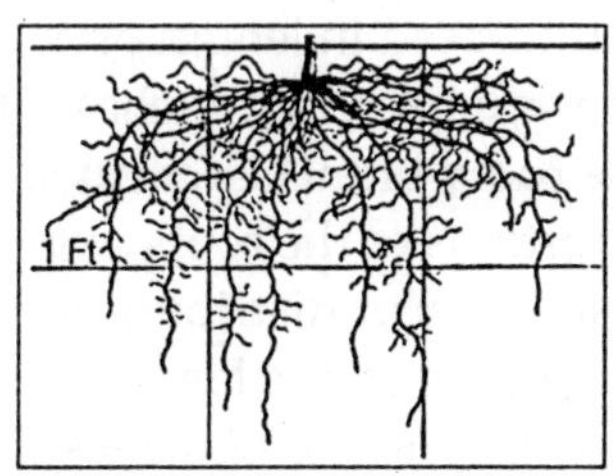

Fig. 5.2 One-half of the Root System of Leek, 3 Months Old.

Maturing Plants

The leek plants were in fine condition and still growing vigorously on July 26 at the final examination. Each had about a dozen leaves, the largest being 2 to 2.5 feet long and 1.5 inches wide at the base.

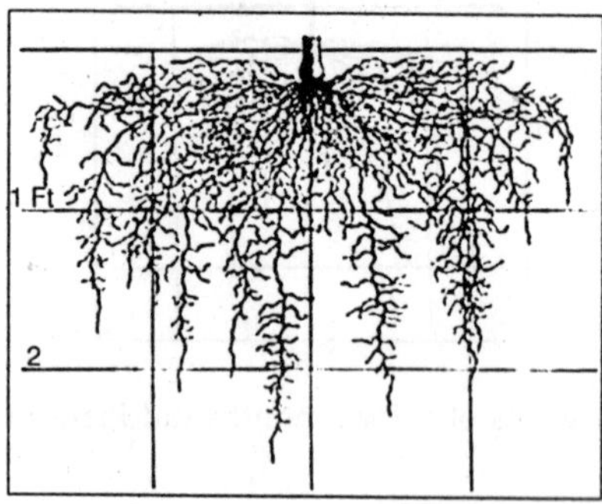

Fig. 5.3 Leek late in July of the First Season's Growth. The Root System was Still Developing Vigorously and Would Probably have Extended much Deeper.

A typical plant had 119 fibrous roots extending in all directions from the base of the stalk from horizontally to vertically downward. Many of these roots were only 1 millimeter in diameter but the larger ones were 1.5 millimeters thick, at least throughout the first foot of their course. The general direction of growth may be seen at a glance where it may be noted that the lateral spread had been increased to 21 inches and the maximum depth to 30 inches.

Tiny laterals usually began to appear about 2 inches from the root tips. These increased in length on the older portions of the root, reaching a maximum of 2 inches a foot from the root ends. On some of the roots, branches occurred quite irregularly. Thus considerable portions of the roots were free from laterals. On others an almost uniform distribution of two rootlets per inch was found throughout their length. On the larger roots, within a radius of approximately 12 inches from the base of the plant, larger branches occurred. Frequently, a length of 10 inches was attained by these laterals, which were, however, only rarely rebranched. The number of laterals per inch of main root had not increased but they were longer than before and occurred over a much larger root area.

SUMMARY

Leek has a fibrous root system which consists of 50 to 100 or more main roots. Many of these spread 14 to 21 inches just beneath the soil surface where they may end or, more usually, turn downward. Others pursue an obliquely downward course so that the soil is ramified with roots of mature plants to a depth of 18 to 24 inches. The branches are moderately few with a maximum length of 10 inches but they rarely rebranch.

Compared with the onion of equal age the root system of leek spreads more widely, a fact that may be correlated with the practice of spacing leek plants a little farther apart than onions. Otherwise cultural practices are similar. The top development of leek is also considerably greater than that of the onion. The depth of penetration is much less but this is partly compensated by the much longer laterals and the more thorough distribution of roots in the surface soil. More profuse branching and more thorough occupancy of a wider area of soil have also been found by other investigators. In both species the laterals are only rarely rebranched. To what degree these differences are heritable and to what extent they may be modified by environment await experimental investigation.

GARLIC

Garlic (Allium sativum) is a hardy, perennial, bulbous herb closely allied to the onion. Unlike the onion which usually produces one large bulb, garlic bulbs are composed of several, small, elongated, egg-shaped bulbils, all of which are enclosed in a whitish skin. In propagation these bulbils or cloves are more commonly used than seeds.

Strong cloves were planted Oct. 1 at Norman, Okla., in double rows 6 inches apart. The cloves were spaced 12 inches apart and a distance of 30 inches was left between the double rows.

EARLY DEVELOPMENT

Owing to favourable weather the plants made a rapid growth and by the middle of December the tops consisted of about five leaves per plant. These were approximately 1 inch in diameter and 6 inches long. They had developed about 18 roots per plant. Most of them spread horizontally at a depth of 2.5 to 5 inches, a few grew nearly vertically downward, and the remainder took an intermediate direction of growth. They pursued a rather zigzag course through the soil, often turning sharply or forming a complete coil where a clod or other obstacle was encountered.

A maximum spread of 9 inches and a depth of 11 inches were attained. Roots that did not exceed a length of 8 inches were destitute of laterals. Older and longer roots were furnished with a few branches 1 inch or less in length. These occurred on all but the youngest one-third of the roots but there were only one to two per inch. The main roots were about 1 millimeter in diameter and very tender. The characteristic alliaceous odour could be readily detected to very near the root ends.

GROWTH DURING THE WINTER

On March 1, just after spring growth had begun, a further study was made. Since no freezes of sufficient severity to kill the tops had occurred, the plants grew during the winter. There were five to six well developed leaves, 6 to 8 inches long, per plant.

The number of roots varied from 25 to 50 depending upon the size and vigour of the plant. The lateral spread had been increased to a maximum of 15 inches and the depth to 22 inches. Although some of the horizontal roots still ended at about the same depth as their origin, many, after extending laterally 8 to 12 inches, turned downward 1 to 3 inches.

The direction of growth of the other roots had not changed. All remained white in colour and about 1 millimeter thick. Within 1 foot from the bulb, laterals were numerous. The longest ones extended 1.5 inches, often towards the soil surface. Farther from the base of the plant, they became fewer and shorter. The rapidly growing, shining, white root ends were free from laterals.

MATURE PLANTS

By June 12 the plants were 2 feet high and well developed clusters of sets had formed. The yellowish leaf tips and the general maturity of the plant indicated that they had nearly completed their growth. The group of bulbs at the base of the plants averaged 1.5 inches in diameter. There were now 40 to 60 roots per plant, some of which extended 18 inches laterally; others penetrated to a depth of 2.5 feet. Some of the main roots had grown to 1.5 millimeters in diameter near the plant. Laterals on these, 8 inches long near the bulb, were not infrequent. Many extended upward in such a manner that the shallowest main roots were 2 to 3 inches deep but the surface soil was thoroughly ramified. All of the laterals were longer than formerly and although not abundant on any one root, yet so many main roots ran through the soil that it was well occupied. At this time the roots did not appear to be growing vigorously. As in its earlier stages of development, the root system was very similar to that of the onion and the leek.

An isolated 2-year-old plant had an even more extensive root system which correlated with a better developed top. The latter consisted of 32 stalks. Each stalk had an average of 25 main roots, 1.5 millimeters in diameter. This thickness was maintained throughout the entire length of the white, succulent roots. A lateral spread of 22 inches, 4 inches greater than that previously found, and a depth of 40 inches were ascertained. The roots were, moreover, growing rapidly. The soil was densely filled with a network of roots. The well-developed laterals typically occurred at the rate of two per inch but not infrequently they were lacking for several inches along the root.

SUMMARY

The root system of garlic is very similar to that of the onion and the leek. The number, size, direction of growth, and length of the main fibrous roots are very similar, A maximum lateral spread of 18 inches and a depth of penetration of 2.5 feet are attained during the first year. The rate of occurrence of the simple laterals is also similar to that of the leek and the onion but the branches are somewhat longer than in the onion. The thorough occupancy of the surface soil was as pronounced as that in the growth of the leek and somewhat more so than that in the growth of the onion. This difference may be due to differences in soil environment.

ASPARAGUS

The common garden asparagus (*Asparagus officinalis altilis*) is a perennial plant with much branched, annual, aerial stems which reach heights of 3 to 10 feet. These, like the long, somewhat fleshy, cord-like roots, arise from a rather short, much thickened, branched, and rather woody rootstock or crown which lies in a horizontal position a few inches below the soil surface. This rootstock grows 1 to 3 inches a year, either at one or both ends; hence, an asparagus plant once established will produce for many years. Asparagus is one of the most popular of the perennial vegetable crops. It is grown on a vast scale commercially and an asparagus bed is to be found in most home gardens. The plants are grown for their thick, soft, young shoots which appear early in the spring.

EARLY DEVELOPMENT

The early growth of asparagus has been thoroughly studied in California. In the growth of the seedling, the single, primary root takes a direct course downward, developing numerous thread-like lateral rootlets. The chief function of this primary root with its laterals is absorption. It seldom attains a length of more than 5 or 6 inches. It is much more slender and fibrous than the storage roots which develop later. The single primary shoot takes a direct course upward and upon reaching the light develops a few side branches and leaves. This primary shoot seldom attains a length of more than 4 or 5 inches. Both the primary root and the primary shoot are temporary organs. They wither and die long before the end of the growing season... The primary root and primary shoot attain a length of 3 to 4 inches before connection with the reserve supply of food in the seed is severed. On the very young crown a brown sear may be observed at the point where the absorbing organ was attached (Fig. 5.4).

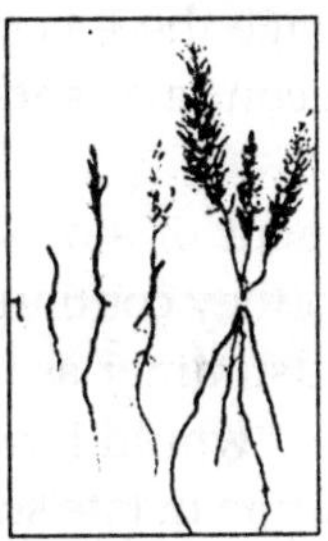

Fig. 5.4 Five Stages in the Development of an Asparagus Seedling.

At the left a very young stage showing the short seminal root and the much shorter seminal shoot, both of which are attached to the seed and are deriving nourishment from the stored food in the endosperm.

Table. Development of Asparagus Plant During First Season 71 Numerical Values are the Average of 20 Plants.

Seeds

	Length of	Length		Maximum		Maximum
Date	Primary	of	Number	Length	Number	Length
of Observ-	Shoot	Primary	Storage	Storage	Secondary	
Secondary						
ation	Above	Root,	Roots	Roots,	Shoots	Shoots,
	Seed,	Inches	Inches	Inches		Inches
Apr. 27	3.0	4.1	0.0	0.0	0.0	0.0
May 19	3.8	5.1	1.6	0.2	1.2	1.1
June 9	4.0	5.2	4.1	5.5	2.1	5.7
June 30	4.2	5.4	6.4	7.5	3.7	7.6
July 24	4.3	5.4	16.1	8.6	5.5	12.1
Aug, 13	...	...	28.3	....	8.1	20.2
Sept.20	...	..	42.0	....	9.0	24.0

In the second and third stages the seed is still attached. In the fourth stage the plant has become independent of stored food in the seed, the seminal shoot has branched slightly, a second shoot has arisen from the crown, and a fleshy root has developed. In the fifth stage there is shown the seminal shoot, two well-developed, secondary shoots and one very short secondary shoot.

The rapid development of storage roots both in number and length is of interest. If they continued their growth in length at the same rate after midsummer as in July, *i.e.*, nearly ½ inch per day, a depth of 3.5 feet would have been attained. Even an accelerated rate of elongation in late summer is not improbable, since the shoots are large enough to furnish much food. Although the plants produced but a single secondary shoot after Aug. 13, there was, nevertheless, a considerable increase in the number of storage roots and in the size of the crown. The storage roots are clothed with fine, fibrous, often unbranched, absorptive roots which extend throughout their course. The fleshy roots increase in number, growing out from the sides and under surface of the rootstock. They continue their growth in length from year to year. Frequently, a definite scar may be found where the renewed growth occurred. Thus a very widely spreading and deeply penetrating root system is produced.

MATURE PLANTS

The common garden asparagus was excavated and studied near Lincoln late in June. The plants had been growing in the field about 6 years. The tops were 4 feet tall and the fruits fully grown but green. The black silt loam soil of loessoid origin was about 16 inches deep and underlaid with a very stiff clay. The clay continued to a depth of 3.5 feet below which it gradually gave way to loess. The soil was quite dry, at least to a depth of 9 feet, and made root excavations difficult.

The long, fleshy storage roots arose from a rootstock or crown which, occurred at a depth of about 6 inches. The fine, fibrous, absorptive roots occurred mostly as branches on the fleshy ones. The branched, rather woody rhizomes were about

1 inch thick and frequently many inches in length, the older portions having rotted. On a section of rhizome only 6 inches long, 140 roots were found. These were 4 to 5 millimeters in diameter and extended in all directions except towards the soil surface. It is quite impossible to show all of the roots in a small-scale drawing.

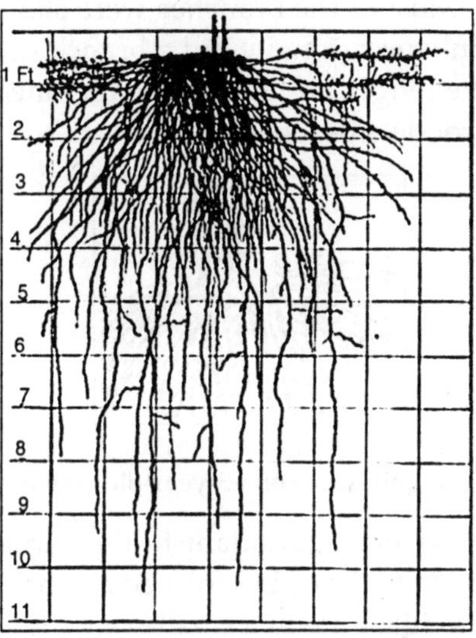

Fig. 5.5 Rootstock and Root System of a 6-year-old Asparagus Plant. Not all of the Very Numerous Main Roots are Shown.

Some of the roots radiated horizontally from the rhizome 4 to 6 feet or more, at depths of only 3 to 8 inches. Although they were 4 to 5 millimeters in diameter, they gradually tapered towards their ends where they were only 2 millimeters thick. They were much more branched near their extremities than close to their origin. For example, in the first 2 to 3 feet of their. course, three to five rootlets per inch about 0.3 inch long and quite unbranched, were found. But on the distal half there were usually five roots per inch, most of which were rebranched. These roots were from 1 to 4 inches. long and formed quite a network in the surface soil. They were rebranched at the rate of two to five rootlets per inch, these secondary branches varying from 0.1 to 1 inch in length. In addition to the horizontal, widely spreading roots, numerous others pursued an obliquely ouitward and downward course before penetrating straight downward. Still others ran almost vertically downward. They showed many kinks and curves in penetrating the clay soil layer. These roots were rather poorly branched at the rate of no laterals to five per inch. Most of the branches were short, 0.2 to 1 inch in length, and only occasionally did a branch several inches in length occur. The longer branches were rebranched at the rate of two to three rootlets per inch.

Fig. 5.6 Crown and Roots of a 3-year-old Asparagus Plant.

Branching was most abundant in the surface foot of soil where the white tender branches formed a dense network. This is very difficult to illustrate in a small-scale drawing. In fact, it appeared that most of the absorption was carried on in the surface 3 feet of soil and especially in the surface foot. The working depth, however, as indicated by the abundance of roots, occurred at 4.5 feet, although some large roots extended deeper, and the maximum penetration was 10.5 feet.

Between 3 and 4.5 feet, roots with a diameter of 1 to 2 millimeters were abundant and a number of the larger roots (5 millimeters thick) ended here. The thicker roots were somewhat better branched than the finer ones, the branches were longer and more of them were rebranched. Below the working depth roots were not numerous. They were 1 to 2 millimeters thick, 0.2 to 1 inch long, and occurred at the rate of no roots to five per inch. Like other short branches on the main roots, they were simple and mostly horizontal but sometimes quite kinky. Occasionally, a rebranched lateral 3 to 5 inches long was found.

Nearer their ends, *i.e.*, below 9 feet, the main roots were only 1 millimeter or less in diameter. The smaller ones had few or no branches, the larger ones had fewer and shorter branches than those just described. The roots at this depth were usually found in earthworm burrows or in the holes formed upon the decay of previous roots.

Frequently, they ran horizontally for a number of inches along these lines of least mechanical resistance in the richer and better aerated soil. Numerous cases were observed where the main roots had ceased elongating at various depths, especially those growing in the stiff clay. Then branches had arisen just above the large root cap and extended for several feet into the soil. These branches had a much smaller diameter than the main roots. The latter, arising from the rhizome, tapered very little towards their ends.

The finer roots were white in colour, fleshy, and very brittle, thus being very difficult to excavate from the rather dry soil. Dead roots in all stages of decomposition were found, occasionally between depths of 8 and 10 feet. They were especially abundant in the surface soil, often consisting of the corky exterior and a small brown or black stele in the center, the food-stored, watery portion of the cortex decaying rapidly.

SUMMARY

Asparagus is a perennial plant with a very extensive root system, which arises from a thick rootstock which grows from year to year. The primary root is only a few inches long and is short lived. It is soon replaced by the thick, long, storage roots which are clothed with short, absorbing laterals. The root system, under favourable conditions, may extend to a depth of 3 feet or more during the first season of growth. With the growth of the rhizome, the roots greatly increase in number. Many radiate into the surface soil 4 to 6 feet even in compact silt loam. Others grow outward and downward and then pursue a vertically downward course. Finally, numerous roots grow almost vertically downward. Thus a very large soil volume is occupied to a working depth of 4.5 feet. Some roots are 9 to 10.5 feet long.

Branching is, however, most abundant, and undoubtedly absorption is very active in the surface foot of soil.

HABIT OF UNDERGROUND PARTS IN RELATION TO CULTURAL PRACTICE

A few of the more important ways in which the roots and rhizomes of asparagus are closely related to the common practices in the successful production of the crop will be briefly discussed.

Soil Preference

That asparagus grows well in almost any kind of soil is indicated by its success as a weed when it escapes from cultivation. A consideration of its root and rhizome habit makes clear why a deep, loose, light type of soil is best. Open, porous soils permit of easy penetration and elongation of the rather thick storage roots. Such soils, moreover, warm earlier in spring and promote a more rapid, early growth. These soils are well drained, an essential environmental factor for good development, since asparagus roots are very sensitive to an excess supply of water which reduces soil aeration.

Another important factor is that connected with the practice of propagating the plant by resetting the crowns and attached roots from plants grown from seed in the nursery. In light soils the crowns may be dug with a minimum of injury. Heavy soils become packed and it is difficult to dig the crowns without injuring many of the rootstocks and losing a large percentage of the fleshy roots. These roots contain the reserve food supply and loss of them lessens the early growth of the plant.

Transplanting

The preference by practical asparagus growers for 1-year-old plants, rather than more mature ones, is undoubtedly related to the extent of root development. With asparagus, as other plants, transplanting checks growth. If the growing point of a fleshy root is uninjured in transplanting it will continue to elongate. Transplanting, however, permits of the selection of the most vigorous plants and also has certain other advantages, especially that of getting a full stand of the crowns started at a proper depth. In the process of transplanting, the roots are spread out evenly on the bottom of the trench or over a small mound of earth placed thereon and moist soil packed firmly about them.

Sowing the seeds in pots and starting the plants in the greenhouse or hotbed, repotting into larger containers, and finally transplanting into the asparagus bed without root injury should promote vigorous development of the plants and increase the succulent stem development earlier in the life of the plant. 154 Experiments in Pennsylvania, extending through a period of 6 years, have shown that the size of the transplanted roots has a direct relation to subsequent yield. The larger, more vigorous plants bore crops with a 26 per cent greater value than the smaller ones. In the process of transplanting, the roots and crowns should not become desiccated or the resumption of growth will be greatly delayed.

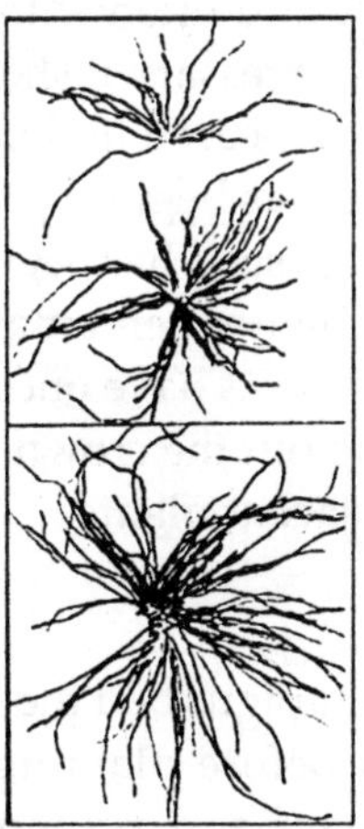

Fig. 5.7 Grades of crowns: "Number 1" (Bottom), "Number 2" (Center), and "Number 3" (Top).

The practice of planting the crowns deeply, although not covered to the extreme depth until the shoots are well up, is directly connected with the root habit. Many of the old fleshy roots die each year and are replaced by new ones. The new roots originate from the rootstock at points somewhat higher than the older ones and thus each year approach nearer the soil surface.

To compensate for this so-called "lifting of the plants," the crowns are placed at least 8 to 10 inches deep. Otherwise in only a few years they would be injured by the process of cultivation. Beds planted too shallowly may last only a few years; plants in deeply planted and properly cultivated and fertilized beds retain their vigour for 10 to 20 years. 151 In all tillage operations constant care must be exercised to prevent injury to the rhizomes and roots.

Growing the Seedlings

In planting asparagus seeds they should be spaced a few inches apart. Where the seeds are dropped in groups, fleshy roots and rootstocks become so interwoven that they are separated with difficulty and often with considerable injury to the plants. If thinning is necessary it should be done before the second aerial shoot has appeared and before the development of the fleshy root system has begun.

After the fleshy root system has begun to develop, the shoot usually breaks at the crown when an attempt is made to pull the plant after the fleshy roots have once started to develop the only way to thin is to dig out the crown, but in doing this there is danger of injuring adjacent plants.

The Permanent Bed

In preparing the asparagus bed the soil is deeply tilled and thoroughly pulverized before planting. This insures a more intimate root-soil contact and promotes vigorous growth. Subsoiling heavier soils and plowing under considerable quantities of manure is also a common practice. This furnishes a better soil structure and at the same time places stores of available nutrients, upon which asparagus draws heavily, in proximity to the growing roots.

Food Storage in Relation to Harvesting

A direct root relation exists in the common practice of permitting the plants to grow 2 full years before the shoots are closely cut. Until the plants have had time to manufacture sufficient food for a good crown and root development, they are materially weakened by removing the tops. Even on older plants the shoots should not be harvested too late in the season but an abundant growth of tops encouraged. The large amount of reserve food required for early spring growth is manufactured by the tops the preceding yeare before their death and stored in the roots and rhizome. The reserve materials stored in the roots in autumn have been shown to be principally sugars. The synthesis of sugar in the tops and its translocation to the roots appear to continue until the tops are killed by frost.

The fertilizing constituents which were stored in the roots over winter appeared to be nearly, if not quite, sufficient for the full development of the succeeding spring crop. The production of young stalks drew most heavily on the sugar contained in the roots, but there was no approach to exhaustion of that constituent.

Relation to Mulch and Humus

The dead tops, if not removed, afford protection for the roots in winter, holding the snow and preventing sudden fluctuation in soil temperature. This is very important in cold climates where the tops are usually supplemented or replaced by a mulch of straw or other material.

When the old roots die and decay, they furnish the soil a considerable amount of humus. This supplements the supply afforded by the decaying tops. Although the plants are widely spaced—usually in rows at least 4 to 6 feet distant and the plants set 1.5 to 2.5 feet apart—yet the extensive root systems fully occupy the soil. The amount of humus is so great that the continued use of manure as a source of humus is sometimes apparently not beneficial. It is thought that the fleshy roots of asparagus may live and function only 1 or 2 years. The constant replacement of older roots by new ones is well known. The older ones then decaying... contribute largely to the humus content of the soil and would seem, therefore, to be a highly important consideration in accounting for the lack of favourable influence of manure on the humus content of asparagus beds. In conclusion it may be said that the rather thorough studies on the root relations of asparagus have thrown much light upon the activities of the whole plant and have led to a far better understanding of its needs and requirements for successful crop production.

A knowledge of the root development of the seedlings has permitted more intelligent practice regarding the preparation of the seed bed, methods of planting, spacing, and thinning; preparation of the permanent bed and the time and manner of transplanting. A study of the older underground parts has thrown much light upon the care, cultivation, and harvesting of the crop and its needs as regards fertilizers, water, mulching, etc. Similar studies on other vegetables will help advance the practice of crop production far beyond the empirical stage.

6

Rhubarb, Beet and Swiss Chard

RHUBARB

Common rhubarb or pieplant (*Rheum rhaponticum*) is a coarse, perennial herb, a new growth arising each year from strong rhizomes. The plant is usually propagated by transplanting portions of the rhizomes and attached roots. It is a cool-season crop, cultivated for its large leafstalks which are available early in the growing season. The plant will withstand the heat of summer, however, and the underground parts are unaffected by severe winter freezing.

MATURE ROOT SYSTEM

Four-year-old plants were excavated in early summer near Lincoln, Neb. The large, leafy tops were well developed. About 16 broadly expanded leaves with blades 15 to 23 inches in length and only slightly less in width grew in clumps of average size. A rich, black, silt loam soil of loessoid origin occurred at a depth of 27 inches. It was underlaid with mellow loess subsoil. A sharp line of demarcation separated these two layers of the soil profile. The upper foot of the deeper layer contained numerous pockets or nodules of calcareous material, the remainder to great depths was fine grained and quite free from concretions. The crowns on selected specimens were about 6 inches in diameter and composed of three to five stems. Each of these short, thick stems was surrounded by dead and partly decayed leaf bases which formed a compact structure extending 5 to 8 inches below the soil surface. From the base of these stems or rhizomes numerous large roots arose.

The largest originated directly beneath the stem cluster and penetrated rather vertically downward; the remainder pursued more horizontal or oblique courses. The largest had a diameter of 3 inches, the others ranged from 1 to 1.5 inches in thickness. In addition a number of smaller roots ranging from 1 to 5 millimeters in diameter arose from the base of the crown and spread horizontally in the surface 12 to 18 inches of soil. These smaller roots branched and rebranched profusely and carried on considerable absorption in the surface soil.

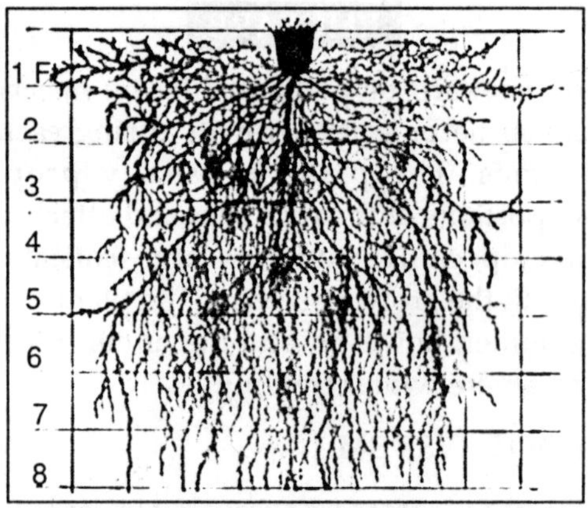

Fig. 6.1 Root System of Rhubarb Four Years Old. Note the Large Number of Sbsorbing Roots Near the Surface of the Soil. Some of the Roots Reached Depths of more than 10 feet.

Six or more strong laterals and numerous smaller ones arose from the main root, originating to a great extent in the second foot of soil. The course of the major branches was vertically or obliquely downward. Many were traced to a depth of 8 feet;

the maximum depth exceeded 10 feet. The thicker portions of the main laterals were nearly always poorly branched with only scattered young rootlets densely covered with root hairs. The characteristic gradual tapering of the fleshy roots to fine extremities, as well as the course pursued by the laterals arising from them, may be best understood by a study of the drawing. The wealth of fine branchlets and the thorough occupancy of the soil within a radius of 3 to 4 feet of the base of the plant and to a depth of 8 feet or more afforded an excellent absorbing system.

A maximum lateral spread of 56 inches was ascertained. The finer roots rebranched at a rate, which although somewhat variable, was strikingly uniform at all depths and in the different layers of soil. Usually 4 to 12 branches per inch of root were found. Many of these were simple and only 0.1 to 1 inch long. Many others were 1 to 4 inches or even more in length and furnished with rootlets at the rate of 2 to 8 per inch. The latter were usually short and simple but not infrequently branched. Sometimes, especially on the coarser branches, 2 to 3 inches of root length were quite free from laterals. Young root systems were better supplied with smaller branches than older ones.

The youngest branches were white but the rest of the root had a characteristic reddish or reddish-brown colour without and a yellow-coloured interior. The roots were brittle and of a rather watery consistency. Even the largest and oldest possessed very little woody tissue. The large branches appeared to increase greatly in diameter with age. The soil beneath the plants was filled to a depth of 8 feet with holes formed by roots now molded away and with many dead roots. These dark-brown to black root remains showed very plainly in the yellowish, loess soil. As the older roots die they are replaced by newer ones. The latter may be readily recognized by the lighter brown or yellow colour.

SUMMARY

The perennial root system of rhubarb is characterized by a thick, fleshy, main root which soon divides into numerous thick branches. These, like the other strong laterals, attenuate gradually and end in very fibrous rootlets. The main roots and their major branches pursue various courses from almost horizontal to nearly vertically downward. Long, slender, much rebranched laterals occur throughout and thoroughly occupy a soil volume with a radius of 3 to 4 feet and extending from the soil surface to a depth of 8 feet.

RELATION OF ROOT SYSTEM TO CULTURAL PRACTICE

A study of the root system explains why rhubarb flourishes in a deep, rich, mellow, well-drained soil rather than in one that is shallow or underlaid with a hardpan. Since it is an early spring crop, a soil that warms rapidly, such as a sandy loam, is best.

A crop should not be harvested until the roots have become well developed and have a reserve supply of food. The practice of encouraging the growth of a leafy plant but of preventing the growth of the flower stalks is directly concerned with the root activities.

The extensive early spring growth of the plant is made at the expense of food stored in the roots during the precedIng year. Hence, sufficient nutrients should be furnished the plants so that they will make a good growth of foliage after the early leaves are harvested and thus be enabled to manufacture this reserve food supply for the roots. In fact, there is usually a direct correlation between a good yield of rhubarb during any year and the growth of the leaves the preceding one. If the coarse flower stalks, often 4 to 6 feet tall, are permitted to grow they utilize much of the food that would otherwise be stored in the roots.

The large supply of reserve foods in the roots is shown by the practice of forcing rhubarb in cellars. In the absence of light, all of the food used by the growing leafstalks comes from the supply already stored in the underground parts before planting them in the forcing bed. The fleshy roots are placed in a shallow layer of moist soil which is provided chiefly for the purpose of supplying water to the plant. In fact root growth is usually very slight. The common practice of spacing the plants 3 to 4 feet apart in rows 4 to 5 feet distant affords sufficient room for root development, although even at this distance there is considerable overlapping in absorbing territory. Where there is a single row, as in most home gardens, closer spacing is permissible. After 4 or 5 years the plants often appear to be "running down." This is thought to be caused by too great a root growth. Growers cut away a part of the roots by plowing closely to the plants in the fall or by spading around the clumps. Thus, the overcrowded mass of roots is reduced and new growth of roots stimulated. This should be done only every 4 or 5 years when the plants seem to require it.

The transpiring surface is very large and the roots must absorb and transport to the leaves large quantities of water. Hence, thorough cultivation to keep out competing weeds and to maintain a soil mulch is necessary for the best development. Because of the proximity of many roots to the soil surface, cultivation should never be deep. Mulching the plants with a heavy application of barnyard manure in the fall of the year protects the roots during the winter and also affects them favourably by enriching the soil. Soils do not freeze so deeply when covered with a mulch, and root activity may be resumed earlier the following spring, especially if a part of the mulch is removed. Rhubarb roots are well fitted for extensive absorption and require a rich soil.

BEET

The garden beet (*Beta vulgaris*) although a biennial is grown as an annual. It is one of the most important of the root crops. Beets are hardy and easily grown and are found not only in a large percentage of market gardens but also in nearly all home gardens. During the first season of growth it accumulates a large amount of food in the fleshy taproot. If the beet is grown a second year, most of the surplus food is used in the production of aerial shoots. These are much branched and leafy and reach a height of 2 to 4 feet. The "beet" itself is largely the fleshy upper portion of a long taproot. The upper part or crown is a very much shortened, fleshy stem, upon the apex of which the leaves are borne. The root proper may be distinguished from the stem or crown by the two opposite, longitudinal rows of secondary roots. Beet seeds of the Edmand's Blood Turnip variety were sown Apr. 24 in drill rows 18 inches distant. Later the seedlings were thinned until they were 5 inches apart.

EARLY DEVELOPMENT

The early root development of the beet is rather rapid. Under very favourable conditions for growth, laterals appear on the upper portion of the long taproot only 7 days after the seed is planted. The first field examination was made June 4 when the plants were 6 weeks old. The tops were 5 inches tall and each plant had 6 to 8 half-grown leaves about 3 inches long and 2.5 inches wide. The total transpiring surface (two sides of the leaves) averaged 1 square foot.

The underground parts were characterized by strong taproots which reached depths of 2.5 feet. The upper portion of the root had already begun to thicken but below 4 inches it was only 1 to 2 millimeters in diameter (Fig. 6.2). Although slightly kinked and curved, the general course was vertically downward.

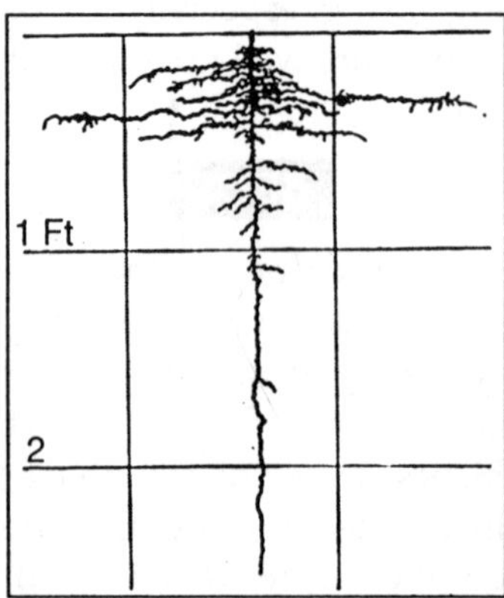

Fig. 6.2 A Garden Beet 6 Weeks.

Rapid growth was indicated by the long, unbranched root ends. Usually more than 100 branches arose in the surface foot of soil. They spread rather horizontally on all sides of the plant. The smaller ones were only 1 inch or less in length and unbranched. The longer ones extended away from the taproot for a distance of a foot or more and were well clothed with short, unbranched laterals. In the second foot the laterals were almost as abundant but only rarely over 0.3 inch in length.

MIDSUMMER GROWTH

A second examination, June 30, revealed marked development. The tops were nearly a foot in average height and each possessed about 16 large leaves. The area presented for photosynthesis and transpiration was very large since the leaf blades were approximately 8 inches long and 5 to 6 inches wide. It had increased to 4.8 square feet.

To supply water and soil nutrients for the rapidly growing tops, the plants had developed a remarkably extensive root system.

The fleshy taproots were now 1.7 inches in greatest diameter but tapered rapidly under the enlarged surface portion so that they were only 3 to 4 millimeters in diameter at a depth of 18 inches. Beyond this depth the taproots, now only 1 millimeter thick, continued their somewhat tortuous but usually vertical course, reaching depths of 60 to 65 inches in the compact subsoil. The taproot was profusely branched, except near the tip, throughout its entire course. The roots originated in two rows on opposite sides of the taproot which is also characteristic of sugar beets. 68 The vigorous growth of the plants was shown by the long, unbranched ends of the taproot and their larger branches. Often 8 to 11 inches of the root ends were entirely unbranched.

From the lower one-third to one-half of the "beet" many, fine, unbranched rootlets arose. These were only 1 to 2.5 inches in length. Within the first foot of soil the taproot gave rise to from 100 to 125 branches. Many of these (approximately one-third) were only 0.5 to 3 inches in length and entirely unbranched. They were very delicate and almost thread-like. The other branches were larger in diameter and varied in length from 4 inches to 3 feet. A few had a horizontal spread of 4.5 feet. Like the smaller laterals they nearly all took a horizontal or slightly downward course. They were branched somewhat irregularly but often at the rate of two to eight laterals per inch. These small branches varied between 0.2 and 1.5 inches in length. Many were furnished with short sublaterals. Thus the surface foot of soil was already well ramified with the roots of this crop.

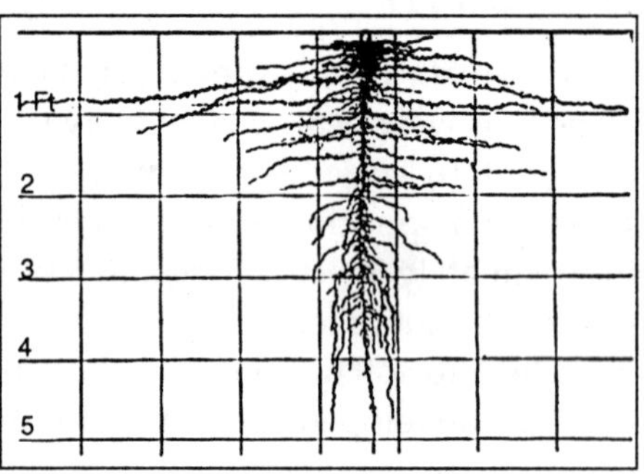

Fig. 6.3 A Beet about 10 Weeks Old.

In the second, third, and even in the fourth foot, branching was scarcely less pronounced. The fine laterals were shorter, however, and the longer ones had not had time to pursue their horizontal course for a distance greater than 2 feet from the taproot. Although the secondary branches were shorter, the tertiary ones were practically absent except in the second foot. The number of branches still ranged between two and eight per inch. In the upper portion of the third foot the soil structure was very compact. Owing perhaps to the hard soil, branching of the laterals was noticeably less than on the roots above or below this stratum. The tendency of the roots in the fourth foot to spread only a few inches horizontally and then turn downward was very characteristic. Like the taproots, the long unbranched root ends indicated rapid growth. On the younger, rapidly growing roots laterals were very short.

A somewhat cone-shaped volume of soil, with its apex at about the 5-foot level and its base extending 2 feet or more on all sides of the plant, was being drawn upon to furnish water and nutrients for each rapidly developing plant. This volume included about 21 cubic feet of soil and subsoil. The excavation, examination, and comparison of such an extensive root system were somewhat easier than usual because of the fact that all except the smallest roots were fairly tough. All but the youngest were characterized by the presence of anthocyanin in such abundance as to give them the characteristic dark-pink or reddish colour of the "beet" proper.

MATURING PLANTS

At the final examination, Aug. 12, 6 weeks later, the plants had an average height of 11 inches although some were 5 inches taller. Plants of average size possessed 24 large leaves, about one-third of which were dead and 4 to 6 of the younger ones were only partially developed. The 10 to 12 fully grown leaves had approximately 60 square inches each of photosynthetic area. The "beets" were now 3.5 to 4 inches in diameter.

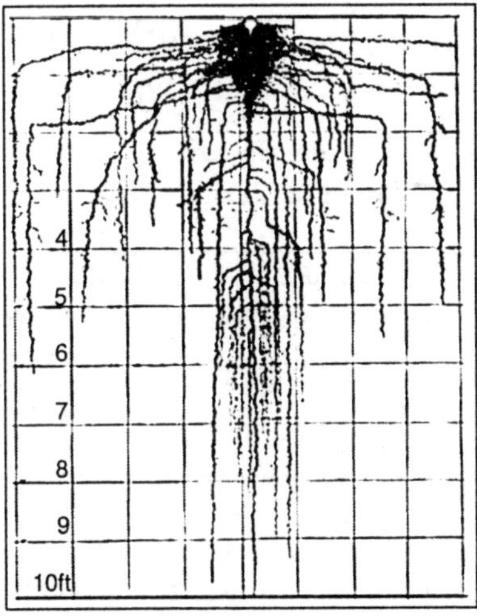

Fig. 6.4 Root System of the Garden Beet on Aug. 12, about 3.5 Months after Planting the Seed.

The taproot had elongated at an average rate of over 1 inch per day. Note the extensive absorbing area deep in the soil. The strong taproots had grown vertically downward, many to the 10-foot level; some reached a maximum depth of 11 feet. Large numbers of the formerly horizontal branches in the 2 feet of surface soil had extended laterally for distances of 2 to 4 feet, and then, turning rather abruptly downward, penetrated to depths of 4 to 6 feet. Moreover, a remarkable development of the deeper portion of the root system had occurred. Here, the downward-growing tendency of the laterals, already shown late in June, was clearly evident. These vertical branches, paralleling the course of the taproot, thoroughly filled the deeper soil, some reaching depths nearly or quite as great as the taproot.

Approximately on the lower half of the "beet" short roots occurred in great profusion. Here, as in the sugar beet, they are confined to two broad rows on opposite sides of the root. As many as 150 roots were counted on a single individual. These roots seldom exceeded 8 inches in length and usually most of them were shorter. Some were unbranched but others were so profusely branched as to form a dense network of rootlets in the surface soil. The abundant branching of the taproots may be seen in Table which illustrates a typical case.

Table. Number of Laterals Arising from 1 Inch of the Taproot at Various Depths

Depth, Inches	Number of Roots Less than 1 Milli-Meter in Diameter	Number of Roots Greater than 1 Milli-Meter in Diameter	Depth, Inches	Number of Roots Less than 1 Milli-Meter in Diameter	Number of Roots Greater than 1 Milli-Meter in Diameter
5	29	4	13	17	2
6	26	5	14	20	1
7	14	4	15 to 2	10 to 15	2
8	12	1	20 to 25	6 to 10	0
9	20	2	25 to 30	6 to 8	1
10	19	3	30 to 35	2 to 6	1
11	25	2	35 to 40	2 to 7	3
12	16	1	40 to 45	3 to 8	3

Many of the smaller branches were hairlike, only 1 inch or less in length, and often free from branchlets. The larger laterals were frequently 1.5 to 2.5 millimeters thick throughout much of their course but all tapered to ½ millimeter or less in diameter several inches from their ends. The rate of branching was quite variable, ranging from 5 to 10 branches per inch. The branchlets were usually 0.1 to 4 inches in length and often quite well furnished with laterals. The course of the laterals may be best seen in the drawing. In the drier and more compact soil layer at a depth of 36 to 45 inches, branching was poorer and few of the roots were over 2 to 4 inches long. But at greater depths from 7 to 10 branches (maximum, 17) normally arose from each inch of the taproot. Some were of large size, ran obliquely, and then turned downward and penetrated deeply. They were clothed with sublaterals, 0.5 to 2 inches long, in numbers similar to that of the taproot. These were poorly rebranched. A working level of 7 feet was found. although numerous roots penetrated more deeply.

The entire root system had the characteristic red or pinkish colour except the root ends which were bright and shiny white. Even these could be immediately identified by the taste as belonging to the beet. Earthworm burrows frequently occurred and seemed most abundant in the deeper soil. Upon entering them the roots branched much more profusely, the branches running parallel to the large laterals in the burrows. This resulted in dense masses of roots of rope-like structure. Some of these branches were also found running horizontally 10 to 12 inches at depths of 7 feet. Although some of the cobwebby mats of roots in the drier, upper soil layers had shriveled and died, examination of the bright, turgid, unbranched root ends clearly showed that the roots were still growing.

ROOT DEVELOPMENT OF SEED BEETS

Beets of the same variety were also grown in Oklahoma. They were left in the soil unprotected during the winter. Only about one-fourth of the plants survived so that they were spaced 1.5 to 2 feet apart in the rows which were 3.5 feet distant. The roots of these surviving plants were dead. Renewed growth of tops began about Mar. 1. Three weeks later 70 to 260 unbranched roots had developed from two rows on opposite sides of the fleshy, somewhat shriveled "beet." These were mostly horizontal or ran obliquely and only slightly downward. They were 0. 1 to 6 inches long.

By Apr. 18 the plant had a crown of leaves 6 to 9 inches high. Aside from a multitude of fine, hairlike rootlets, about 40 were 1 to 2 millimeters in diameter. These extended 6 to 18 inches on all sides of the plant (maximum spread 23 inches); some penetrated obliquely and reached a depth of 22 inches. Many ended in the 14- to 18-inch soil level. The older parts of the larger roots were furnished with laterals 3 to 4 inches long at the rate of four to six pet inch. Many of the smaller roots were also branched, so that quite a network of rootlets filled the soil. These new roots are necessary to supply the demands for water made by the new shoots. They also absorb food materials from the soil which supplement the supply accumulated the preceding year and which thus promote a good growth of tops and an abundant yield of seed.

The plants made a vigorous growth during May and by the end of the month were about 3 feet tall and had five to nine well-branched, leafy stems. In fact they had almost attained their maximum development. The flowering period was nearly passed and fruits had begun to mature.

The root system was very different from that of the first season. About 50 long roots, varying in diameter from 0. 5 to 2 millimeters, arose from the lower portion of the "beet" and from the 2 to 3 inches of the attenuated taproot which survived the winter. None of these roots had major branches but were cord-like in character, maintaining their original diameter almost throughout their course. Many of them grew rather parallel to the soil surface at depths of 3 to 12 inches, some reaching a distance of 37 feet from the base of the plant. Others penetrated outward and obliquely downward to distances of 12 to 24 inches or more and ended in the second foot of mellow, moist, sandy soil. Still others pursued a more vertically downward course.

These extended deepest, the longest to the 35- to 41-inch soil level. The paths of some of the roots were quite tortuous. The root distribution and the degree of branching are shown in Fig. 6.5. The larger roots were rather poorly branched on their older portions near the plant (two to eight rootlets per inch). This part of the soil, however, was thoroughly ramified by finer, wellbranched rootlets. But throughout most of the course of the roots, except near their tips, branches, usually 1 to 4 inches long but mostly unbranched, occurred at the rate of 16 to 20 per inch.

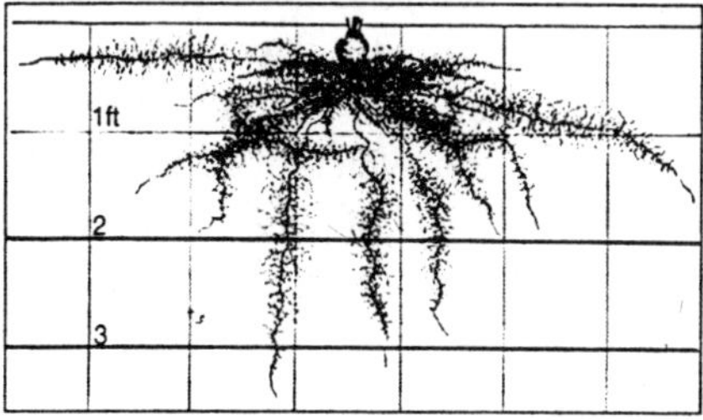

Fig. 6.5 A Portion of the Root System of a "Seed Beet" Reset in the Spring Following the First Season of Growth. The Old Roots had Died and all the Firbous Roots Grew During the Second Season.

Thus the new root system was not only widely spreading but well furnished with efficient absorbing rootlets. This root development is similar to that of "mother sugar beets " used for production of seed. In experiments with these it was found that the lower temperatures of early spring or late fall gave a more profuse and more extensive root development than when the beets were set out during warmer weather. This temperature relation was clearly marked. It is of interest to note that although the roots of the beets were entirely winterkilled, those of the carrot were only partially injured, and the roots of the parsnip lived through the winter unharmed.

SUMMARY

The beet has a very pronounced tap root which grows at an average rate of over 1 inch per day during a period of 3.5 months. By the last of June it is 5 feet long and in August extends to the 10-foot level. Two types of branch roots occur in the first 2 feet of soil. The first is exceedingly numerous, short, and repeatedly rebranched. On mature plants they fill the soil in a cone-shaped volume about 16 inches wide near the surface. (where they grow out in rows on two sides of the "beet") to its apex at a depth of 2 feet. This portion of the root system develops rather late. Strong branches, intermingled with these shorter roots, also extend horizontally, or only slightly obliquely, some to a distance of 4 feet. Turning vertically downward they then 'reach depths of 3 to 6 feet. Below 2 feet branches from the taproot grow out more obliquely and turn downward in such a manner that the lateral spread scarcely exceeds 1 foot. They form, with the taproot, an efficient absorbing system in the deeper soil (2 to 8 feet). All of the main roots are profusely branched but with short laterals. Hence, although the root system is very extensive, the soil volume delimited by it is not fully occupied.

The root system of a "seed beet" consists of 40 to 60 fibrous or cord-like roots running horizontally, obliquely, or rather vertically downward. These are new roots which arise from the lower portion of the "beet" and the short remnant of the attenuated part of the original taproot. All are well clothed with unbranched laterals. They thoroughly ramify a hemispherical soil volume having a radius of about 3 feet.

OTHER INVESTIGATIONS ON BEETS

It has been observed in Russia that during the first 3 weeks after planting the root development is weak. It later becomes accelerated to a point where the taproots grow 1.2 inches per day and the lateral branches elongate at one-half this rate. A root depth of nearly 5 feet and a horizontal spread of nearly 22 inches were attained.

An examination of the Extra Long Dark Blood beet at Geneva, N. Y., showed that the main root was smooth and symmetrical to a depth of 8 inches. Below this it divided into several branches, which were quite thick at first, but rapidly tapered to only a few millimeters in diameter and then to thread-like proportions. One of the longer ones extended 2 feet downward, while horizontal branches, which were mostly shallow in the soil, extended a distance of 2.5 feet. The small fibrous roots seen on the surface of beet roots after they are pulled seem to have very little office, as they penetrate the soil scarcely ½ inch. The feeding roots chiefly proceed from the taproot, below the thickened portion. Fibrous roots from the branches often extend upward apparently to the surface of the ground. The root system of the Eclipse beet, which is a turnip-rooted variety, growing largely above ground, is precisely similar in kind but slightly less extensive. We traced the roots downward about 22 inches, and horizontally a distance of 2 feet.

There has been observed at this station an early variety of beet and also of radish which appeared to root shallower than later ones, although no such differences were found in lettuce and pea.

The rooting habits of Crosby Egyptian beet, an early variety, was also studied at Ithaca, N. Y. They were grown in a fertile gravelly sandy loam soil underlain at a depth of 8 to 12 inches with gravelly sandy loam which became sandier at increasing depths. On plants 12 inches tall and with the fleshy taproot 0.5 to 1 inch in diameter, very few roots had penetrated to a greater depth than 6 inches, although the taproot reached a depth of 15 inches. The space between the 18-inch rows was fairly well filled with roots to the depth of 4 inches. When the plants were nearly fully grown and the beets 2.5 to 3.5 inches in diameter the taproot extended to a depth of 2 feet, and some branches from this to 30 inches.

Small branch roots developed from the taproot throughout its length, some of which were 18 inches long. Near the surface a large number of roots ran almost horizontally and extended from one row to another, there being nearly as many in the centers as near the rows. The lateral roots were small and of nearly the same size throughout their length and had many small branch roots. The greatest development of roots was found in the surface 3 or 4 inches of soil where they extended from one row to another, there. being nearly as many in the centers as near the rows, and many were so near the surface that any kind of cultivation would result in some destruction.

ROOT HABIT COMPARED WITH THE SUGAR BEET

The sugar beet is one of the forms of the complex species, *Beta vulgaris,* quite similar to the garden beet in root habit. Like the garden beet it develops a strong, deep taproot, a surface root system of profuse, much-branched, widely spreading laterals, and a more vertically penetrating but extensive system of branches which ramify the deeper soil. Extensive experiments in fine sandy loam soil at Greeley, Colo., have shown that the root system is very susceptible to modifications brought about in the soil environment by variations in the water content, fertility, etc.

It seems prob able that the garden beet would respond in a similar manner. For example, the Kleinwanzelebener variety of sugar beet in dry soil had a smaller taproot which pursued a more tortuous course, did not penetrate so deeply, and was branched more nearly to the tip than similar plants in moist soil. The larger, deeper-seated branches turned downward rather abruptly, reaching depths of 3 to 4 feet. Branching was more profuse throughout. Development of the surface absorbing system may be greatly delayed, although it branches more profusely and may extend even more widely when the soil becomes moist. The root systems of mature plants grown under irrigation and other very favourable conditions were less extensive than that of the garden beet. Depths of 5 to 6 feet were attained and the lateral spread seldom exceeded 18 inches.

Studies in Germany show that not only the sugar beet but also mangels and certain other closely related, fleshy rooted forms are very similar in habit to the garden beet just described.

ROOT HABITS IN RELATION TO CULTURAL PRACTICE

The very deeply penetrating root system of the beet explains why, like other root crops, it thrives best in a deep, friable, well-drained, but moist soil. A consideration of the very fleshy portion of the taproot makes clear why a deep, mellow, easily moved soil is essential for a proper development of the beet. In heavy soil the beets are likely to be unsymmetrical in form. Hence, in preparing the seed bed the soil should be well pulverized, loose, and smooth, but not so loose that it quickly dries. A heavy soil is less satisfactory. It becomes hard and cracks or if kept wet it puddles, conditions very unfavourable for the germination of seed. The actual depth of cultivation varies, of course, with the nature of the soil and the previous depth of cultivation. Poor soil preparation is likely to result in an inferior stand of plants, regardless of the high quality of the seed sown. No amount of later cultivation will compensate for carelessness in the preparation of a seed bed. Early and frequent tillage not only keeps down weeds but also prevents the formation of a soil crust and the subsequent difficulty in the emergence of the shoot.

It promotes good aeration. Since each fruit of the beet contains more than one seed, the plants come up in clumps and must be thinned. The time and manner of thinning is closely related to root injury. Late thinning greatly disturbs the roots of even the more vigorous plants left to mature. The widely spreading and deeply penetrating root systems are undoubtedly an important factor in competition and the resultant reduction of yield where the crop is too thickly grown.

A study of the root system shows that the beet used in this investigation did not depend largely on the surface 4 to 6 inches of the soil for its water and nutrient supplies. A comparison of the soil-moisture data in Table shows that cabbage reduced the water content in the surface foot of the soil to a much greater degree than did the beet. At greater depths, however, the soil in the beet plats was usually drier. It would seem that cultivation of the soil for the purpose of retaining moisture should have some advantage over removing the weeds by scraping. At Ithaca, N. Y., an average gain in yield of 4.25 per cent as a result of a soil mulch has been reported for an early beet which rooted extensively in the surface soil. 159a The large tops, of course, would more or less thoroughly shade the soil and thus retard surface evaporation. Only a few inches of surface soil, except that close to the plant, were, under the conditions of growth described, unoccupied by roots. The lack of superficial roots, except those very close to the plant, permits of deeper cultivation without root injury than is possible in the case of many garden crops. Some varieties, however, and perhaps all varieties under certain conditions, fill the surface soil also with a network of roots. Likewise, a practice of getting manure and other fertilizer worked well into the deeper soil would seem justifiable. This would not only place the manure more easily within the reach of the roots but also at the same time improve the physical condition of the soil.

SWISS CHARD

Chard (*Beta vulgaris cicla*) is a biennial plant with a strong taproot and large leaves clustered on a short stem near the soil surface. It is really a foliage beet and is grown as an annual. Cultivation has changed its habit of growth so that leaves instead of roots are best developed. The leaves, for which it is grown, have large, fleshy leafstalks and broad, crisp blades. It withstands the heat of summer better than most crops grown for greens. The leaves are prepared like those of spinach by boiling and are canned for winter use. The petioles and midribs are frequently cooked and served like asparagus.

Seed was sown at Norman, Okla., Apr. 19, in rows 3.5 feet apart so as to permit cultivation with a horse-drawn harrow. The seed readily germinated in the warm, moist soil and by May 14 thrifty seedlings, 3 to 5 inches high, were abundant. They were thinned to 8 inches apart in the row.

EARLY DEVELOPMENT

The root system at this time consisted of a taproot which branched profusely in the first 6 inches of soil, giving rise to both large and small laterals at the rate of about eight per inch. These extended outward and usually slightly downward, although a few ran rather sharply obliquely downward. The greatest lateral spread was 14 inches and a depth of 19 inches was attained. The laterals and the taproot below 6 inches bore unbranched rootlets for the most part; only a few of the stronger ones were rebranched. Secondary laterals near the taproot were about 2 inches long but rapidly decreased in length on the younger portions of the main branches.

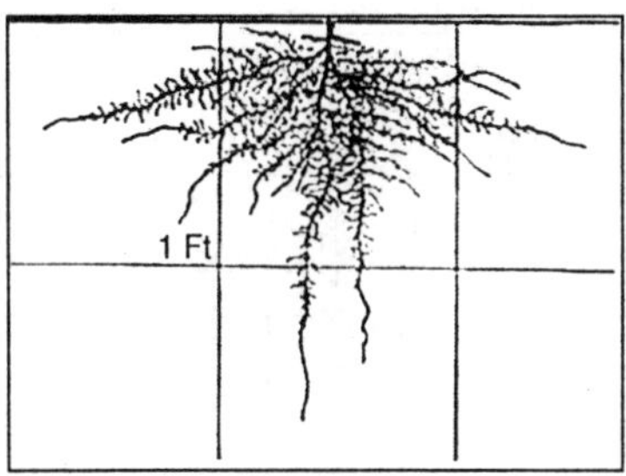

Fig. 6.6 Root System of Swiss Chard 25 Days Old.

HALF-GROWN PLANTS

The seedlings had developed into vigorous, rapidly growing plants by June 7. They were 1 foot tall and each had 8 to 10 leaves. The larger leaf blades were 4 inches wide.

The strong taproot and some of its larger branches had become somewhat fleshy. The gradually tapering taproot penetrated, with small irregularities, directly downward to a depth of 45 inches. As before, the largest branches were on the first 6 inches of the taproot. Many of these had a horizontal course throughout and branched very profusely in the moist, mellow, surface soil. A maximum spread of 2.5 feet was attained. Other laterals turned downward within 1 foot from their origin and pursued a generally vertical course parallel to that of the taproot to depths of 2 to nearly 3 feet. The portion of the taproot below 6 inches was very well branched, some of the branches exceeding 1 foot in length. Their direction of growth varied from horizontally outward to vertically downward. The larger roots of the entire plant were well supplied with absorbing laterals at the rate of 6 to 12 per inch. Some of the older of these unbranched rootlets were 4 inches long. On the first few inches of the taproot and on some of its larger branches the absorbing laterals were beginning to die. Otherwise the conical mass of soil, 5 feet broad and 3.5 feet deep, was well filled throughout with an actively absorbing root network. The plants were again thinned, this time to 2 feet apart in the row. This reduced competition and permitted better growth of the individuals.

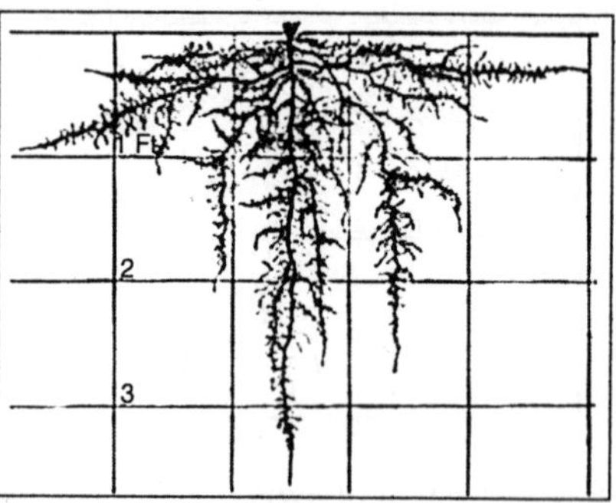

Fig. 6.7 Half-grown Plant of Swiss Chard.

MIDSUMMER DEVELOPMENT

By July 28 the tall, branching crowns were thickly clothed with a dense foliage. As many as 25 leaves, 1.5 to 2 feet long, were common on individual plants.

The taproots were about 2 inches thick. They had increased greatly not only in extent but also in complexity of branching. With characteristic short turns, one taproot penetrated vertically downward, branching freely, to a depth of 4 feet. Here it divided into two rather equal parts but continued its course to a depth of nearly 7 feet.

Most of the taproots and their longest, stronger branches did not extend beyond 6 feet, but One was traced to the 7.5-foot level. Accompanied by strong branches, descending parallel with it, the taproot and its profuse laterals occupied a soil volume which, below the second foot, was about 18 inches in diameter and extended nearly to the 6-foot level.

The part of the root system in the surface 2 feet had similarly increased. Large, horizontal branches, 10 to 15 millimeters in diameter at their origin, extended widely in the surface foot of soil. Running horizontally at a depth of 4 to 8 inches, they extended to distances of 3 to 4 feet (maximum spread, 53 inches) from the base of the plant.

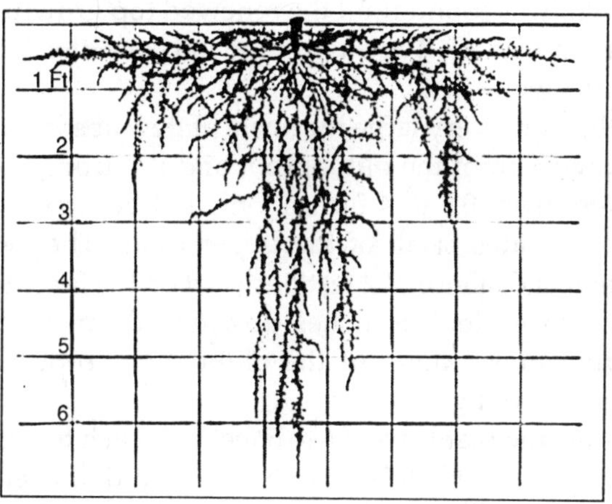

Fig. 6.8 Midsummer Root Development of Swiss Chard. Only one-half of the Root System is Shown.

Although the small absorbing laterals in the first 2 to 2.5 feet of their course had mostly withered, at irregular intervals (often 0.5 to 2 inches or more) they gave rise to larger branches. Similar branches, of course, occurred farther out on these roots, intermixed with the simple laterals. They grew in all directions, often upward near the soil surface. The longest ones, some of which penetrated almost vertically downward, were 2 to 2.5 feet in extent. These and their major branches provided a network of roots throughout the surface 2 feet of soil upon which small laterals grew at the rate of 12 per inch. Thus the root system occupied a large soil volume and was well fitted to draw upon the abundant supplies of water and soil nutrients.

ROOT DEVELOPMENT OF SEED PLANTS

The degree to which the old root system of biennial or winter annual vegetable crops deteriorates and to which a new growth of roots occurs is a subject which deserves thorough investigation. The following observations are indicative of the behaviour of Swiss chard: Plants grown during 1925 were left undisturbed in the field at the end of the growing season, and the root behaviour was examined during winter and spring. No root growth was found until the renewed top growth of spring began about Mar. 15. The environmental conditions during the period are given.

At this time Very large numbers of new absorbing rootlets were growing from the old taproot and its stronger branches. These were from 0.5 to 4 inches long and occurred in clumps on opposite sides of all of the larger roots. They were very abundant to a distance of 2 feet outward on the large horizontal roots and extended 18 inches downward on the vertically descending ones. Many of the older, finer roots were also showing renewed growth.

By June the plants were 4 to 5 feet tall. Each had from three to six well-branched stalks 1 to 3 inches in diameter. The soil within 2 feet of the plant, had been very thoroughly exhausted of its available water. The dense growth of roots just described was dead. But at distances greater than 2 feet from the taproot, the horizontal roots were well supplied with an abundance of fine laterals which grew from the finer branches of the old root system.

The root system as a whole was no more extensive in regard to the volume of soil occupied than at the March examination.. But branching was more profuse near the extremities of both vertical and horizontal roots.

ROOT DEVELOPMENT COMPARED WITH BEET

Compared to its close relative, the garden beet, the root system lacked the network of fine absorbing laterals originating from the taproot in the surface 12 to 18 inches of soil. Neither did the widely spreading main laterals, arising in the surface foot, show the habit of finally turning downward and penetrating deeply. Instead they gave rise to many, large, downwardly growing branches, a root habit quite different from that of the beet. Moreover, long branches, originating from the taproot in the deep soil, were not found, although they are very characteristic in the garden beet. In these and other minor respects the root habits are quite different.

SUMMARY

The root system of Swiss chard is characterized by a strong taproot and many strong lateral branches all of which originate in the first foot of soil. It makes a vigorous growth and early in June occupies a conical mass of soil 5 feet in diameter and 3.5 feet deep. Mature taproots are 2 inches thick, but taper rapidly, and penetrate to depths of 6 to 7 feet. The strong, widely spreading, horizontal branches, with their long and profusely rebranched laterals, thoroughly occupy the soil from near its surface to a depth of 2 feet and throughout a radius of 3.5 feet from the plant.

Almost vertically descending laterals, paralleling the course of the taproot, and like it well clothed with branches, ramify a soil volume about 18 inches wide to a depth of 6 feet. Thus the root is well fitted for absorption both in the surface and in the deep soil. During the second year the plant increases its absorbing surface by a dense growth of rootlets on the older portion of the root system. This is further augmented by the renewed growth of many of the finer rootlets already formed on both the surface and deeper portions of the root system.

OTHER INVESTIGATIONS ON SWISS CHARD

It has been found at Geneva, N. Y., that the root system of Swiss chard is decidedly more extensive than that of the garden beet. In September, the roots of Beck's Sea Kale chard were traced horizontally to a distance of 3.5 feet. At a depth of 2 feet the taproot was still 4 to 5 millimeters thick. The taproot and larger branches were thick and fleshy near the surface, the former regularly tapering as it extended downward, giving rise to branches on all sides. Some of the latter were ¾ inch in diameter, the larger ones starting about 4 inches below the surface. Fibrous roots were numerous in the upper layers of the soil. The chard is a plant of the beet family in which the foliage instead of the root has been developed through selection. It is interesting to observe that with a decided increase of foliage over the common garden beet, we have a corresponding extension of roots.

ROOT DEVELOPMENT IN RELATION TO CULTURAL PRACTICE

The vigorous development of the very extensive root system and perhaps its ability to adapt itself, like the sugar beet, to different environments, may account for the fact that chard is not as exacting in soil requirements as most other vegetable crops. It is a hardy, vigorous plant, and with a little protection will live through the winter. Any good garden soil is satisfactory for this easily grown plant.

It would seem that in growing chard, as well as most deeply rooting plants, that a deeply rooted crop like clover preceding it would be most favourable as a green-manure crop to supply humus. The roots of the clovers extend deeply and have a very beneficial effect upon the soils to a greater depth than most green-manure crops. The effects of roots of legumes upon their death and decay in loosening and aerating the deeper soil are very great. After clover and alfalfa have been grown in it, the soil is quite filled with root channels which greatly modify its structure and promote aeration.

EFFECT ON SOIL STRUCTURE

Extensive, root systems like those of chard and beet must exert a pronounced effect upon soil structure. It seems that with the usual spacing of these plants, the soil would be thoroughly ramified with roots. Investigations on the influence of plant roots on the structure of the soil show that in loose soils a very small percentage of the soil spaces is filled by the stronger roots so that no essential decrease in the original mellowness from this source occurs. But in compact soils roots may to a certain degree improve the structure and thus increase production. In compact stiff soil without granular structure, the loosening process is aided, to the benefit of plant growth, by the mechanical action of roots and by a strong modification of the moisture conditions.

The beneficial combination of self-loosening and root action explains the frequent permanent improvement of the soil structure under the continued influence of roots, as in grasslands, and also the prevention of the permanent puddling of the soil by rain. It has been found that a marked surface spreading of roots has a beneficial influence upon the penetration and movement of water in the soil. Water movement is much more rapid, although lateral percolation and loss of water through evaporation is retarded. Where the soil is occupied by plants throughout the year, the effects are most marked.

7

Spinach, Cabbage and Cauliflower

SPINACH

Spinach (*Spinacia oleracea*) is an erect, smooth, annual herb of rapid growth, related to the beet. It is the most important of the potherbs or greens grown in this country. It is a hardy plant, and the foliage for which it is grown develops rapidly. Early in the season numerous large leaves are crowded on the short stem just above the soil surface. In the North it is grown as an early spring and late fall crop but in the south mainly as a winter crop. During the summer the plant produces a flower stalk 2 to 3 feet in height and develops seed.

Seed of the Curled Savoy (variety *inermis*) was planted Apr. 10 in drills 12 inches apart. After the plants were well established, they were thinned to 4 inches distant in the row.

EARLY DEVELOPMENT

At the initial examination, May 23, the plants had 10 to 12 leaves each. About seven of the largest had blades 2 to 4 inches long and 1.5 to 3 inches wide. The transpiring surface averaged 134 square inches, exclusive of the yellowed cotyledons which were still on the plants. Spinach is characterized by a strong, vertically and deeply penetrating taproot. The root near the soil

Secondary branches commonly varied from 0.5 to 1 inch in length but occasionally branches 2 to 8 inches long occurred. These were clothed with laterals at the same rate as the main branches which was about seven per inch. Short branches of the third order frequently occurred in abundance. Moreover, there were many more laterals on this portion of the taproot than at the earlier examination.

Figure 7.2 shows that the widely spreading laterals (maximum extent, 14 inches), some of which descended to the 2-foot level or beyond, were also well branched. Branches on the taproot below 8 inches were, with few exceptions, quite horizontal but did not spread widely. Below 1.5 feet they frequently occurred in groups, sometimes being confined to one side of the root for a short distance. Frequently as many as 12 arose from a single inch of taproot. The unbranched condition of the younger roots, as well as the absence of laterals from the main root ends, indicated that growth was proceeding rapidly.

MATURE PLANTS

A final examination was made July 10 on maturity of the plants. They had stems over ½ inch thick and 2 feet high. All the basal leaves and many of the stem leaves were dry. The plants had blossomed and set fruit.

The taproot had a diameter of over 1 inch near the soil surface but tapered gradually to 1 millimeter in thickness at the 15-inch level. It reached a depth of 6 feet. In the first 6 inches of its course 40 to 46 laterals took their origin. Nearly half of these were large ones. On the deeper half of the first foot of taproot, however, only 18 to 20 laterals were found and all but one or two of these were relatively small. In fact no large branches occurred below the 10-inch soil level. But to a depth of 2 feet small branches were found in great profusion—as many. as 80 on the second foot of taproot alone. These varied in length from 0.5 to 3 inches. The longest were rebranched with short rootlets. At still greater depths branches arose at the same rate as at the preceding examination, about 8 to 12 per inch. These had increased somewhat in length; most of the rootlets still pursued a horizontal course; and sublaterals were more abundant. The widely spreading laterals which originated in the surface 6 inches of soil now reached depths of 3 to 4 feet. Also some of the larger branches originating at greater depths had extended outward and then turned downward so that the soil around the taproot was more thoroughly occupied. Many of the root ends were decayed. Branches on these main laterals 5 to 10 inches in length were not uncommon. The rate of branching was 6 to 18 laterals per inch and the length varied usually from 0.2 to 3 inches. Branchlets of the third order were abundant.

SUMMARY

Spinach has a pronounced taproot which grows rapidly, penetrates deeply, and gives rise to major branches only in the 6 to 10 inches of surface soil. These spread rather horizontally, usually a foot or less, and then, turning downward, penetrate to depths of 3 to 4 feet. Branching on the first foot of the taproot is very profuse; the finer branches with the laterals from the main roots quite fill the soil volume, the upper surface of which is early blocked out. The taproot, below the first foot, is clothed with relatively short but numerous branches. These add considerably to the absorbing surface in the deeper soil. A depth of 6 feet is attained.

OTHER INVESTIGATIONS ON SPINACH

The deepest growing root extended downward about 2 feet and the longest horizontal roots reached about 18 inches. The feeding roots seemed chiefly to lie at a depth of about 6 inches, though many fibrous roots rose upward to within 2 inches of the surface. The root was a thickened taproot to the depth of 4 inches, below which it divided into many branches of varying length and thickness. Allowing for differences in soil structure, etc., it would seem that the method employed of washing away the soil failed in revealing the entire root extent.

ROOT DEVELOPMENT IN RELATION TO CULTURAL PRACTICE

The extensive root system, rapid growth, and great

transpiring area are clearly correlated with the best soil combination for the growth of spinach, *viz.*, one that is rich and well drained but constantly moist. Hence, the practice on flat lands of plowing the soil into low, flat beds. The extra drainage afforded also keeps the soil from "heaving" and thus tearing the roots of the winter crop.

The plant did not develop a shallow portion to its root system as do many vegetable crops and it is thus less likely to have its roots disturbed by cultivation. The lack of shallow root development may have been due in part, however, to a dry surface soil. In more moist soil many roots might pursue a course nearer the soil surface. The extensive occupation of the surface 2 to 8 inches of soil suggests not only a thorough preparation of this portion of the substratum in regard to humus content and good tilth but also the most effective depth at which to place the fertilizer. The plant must make a quick growth to be crisp and tender and abundant soluble nitrate compounds within reach of the roots promote development of foliage. The usual spacing of the plants, about 6 inches apart in rows 8 to 14 inches distant, results in considerable root overlapping and a complete occupancy of the soil. Root competition is ameliorated to a great extent, however, by well-prepared, very rich soil in which the moisture is conserved by weed eradication and by the preservation of a surface mulch.

CABBAGE

Cabbage (*Brassica oleracea capitata*) is a hardy, biennial plant, although grown as an annual crop. It is one of the most important of vegetables. It is a cool-weather crop, properly hardened plants being able to resist temperatures much below freezing. In the South it makes its growth mainly in the spring or fall. Like the other cole crops, cabbage is grown for its vegetative aboveground parts. The common cabbage, during the first year of its growth, produces a short stem which terminates in a large bud composed of thick, overlapping, smooth leaves, the whole structure being known as the "head." Different varieties are adaptable to a wide climatic range, and it is grown almost throughout the United States. In the North, especially, the early cabbage is usually started under protection and transplanted into the field.

EARLY DEVELOPMENT

The first examination was made June 4. The plants were about 5 inches tall and each usually possessed 20 to 25 leaves. The larger leaf blades, about 10 in number, averaged 4.5 inches both in length and width; the remainder only 2.5 inches in these dimensions. Although the plants were still small, the transpiring area, including both upper and lower leaf surfaces, was already only slightly less than 3 square feet. In all of the several plants examined a main root arose from the base of the enlarged underground part (the taproot having been destroyed in transplanting) and ran in a somewhat tortuous downward course. These roots reached maximum depths of 38 inches. Most of the laterals ran outward and downward, often at an angle of approximately 45 degrees or more from the perpendicular, to depths of 6 to 8 inches. At this depth many pursued a rather horizontal course in the second 6 inches of soil. They frequently reached a lateral spread of 3 feet on all sides of the plant. A few continued their obliquely downward course, ending 1.5 to 2.5 feet from the main vertical root at depths of 12 to 24 inches.

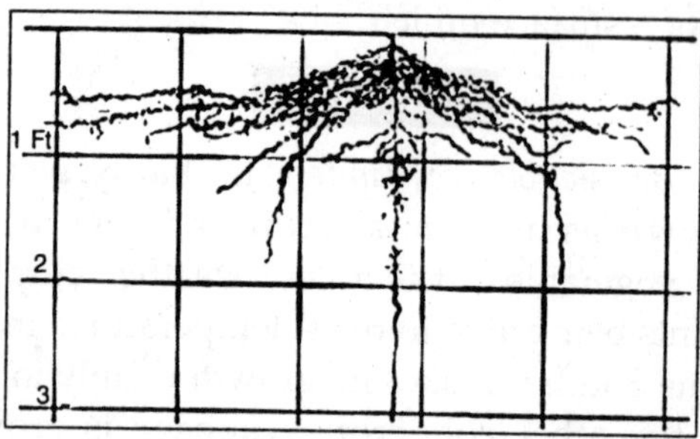

Fig. 7.3 Root System of Copenhagen Market Cabbage 55 Days After Transplanting into the Field.

Few branches arose in the surface 2 to 3 inches of soil. The main lateral roots, usually 22 to 28 in number, were only fairly well clothed with branches. Sometimes an inch of root had only two to four branches but usually laterals 0.5 to 1.5 inches in length occurred at the rate of 4 to 7 per inch. These were unbranched.

The upper portion of the main vertical root had a few rebranched laterals 3 to 9 inches long. Below 12 inches they seldom exceeded 3 inches in length and the deepest portion was quite unbranched. Since practically no roots occurred in the surface 6 inches of soil, except near the plant, cultivation at this time would have resulted in a minimum of root injury.

MIDSUMMER GROWTH

Twenty days later, June 24, the cabbage was again examined. The plants were 1 foot high and had a total leaf spread of 26 inches. The transpiring area had increased to 19 square feet. The heads were 4 to 5 inches in diameter.

During the 20-day period a remarkable development of the root system had occurred. The main vertical roots now reached depths of 4.5 to nearly 5 feet. The widely extending main lateral roots ran obliquely outward and downward and at a depth of 12 inches were 2.5 feet horizontally from the base of the plant. A maximum lateral spread of 41 inches was attained near the 3-foot level. These long roots reached depths of 3 feet or more.

Within the soil volume thus delimited many roots arose from the enlarged underground part. They pursued such varied courses that the soil was quite filled with them and their branches to a depth of at least 3 feet. Reference to Fig. 7.4 shows that many of the rather horizontal laterals of the earlier stage had grown downward and that new roots had occupied the soil directly beneath the plant.

Practically no roots arose from the upper part of the enlarged basal portion of the plant (which had a diameter of nearly 1 inch) but from its terminal part it gave rise to 30 to 38 roots. Most of these ranged from 3 to 6 millimeters in diameter, a few were smaller. These long, cord-like roots tapered gradually so that below 18 inches none were more than I millimeter in diameter and usually less.

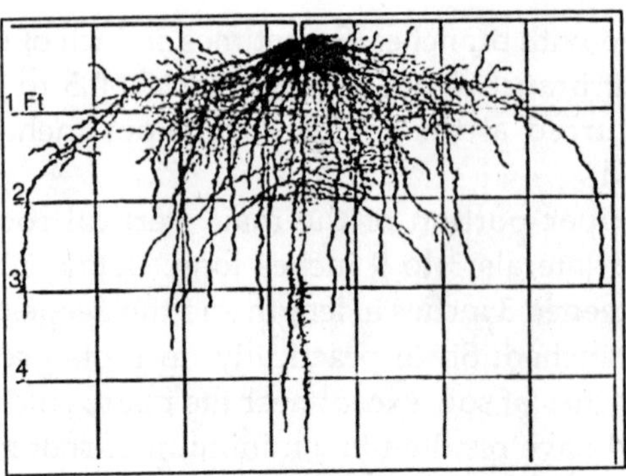

Fig. 7.4 Cabbage Excavated 20 Days later.

Branching, especially in the first foot of soil, was very profuse. From a large number of counts it was found that the number of branches per inch on the first foot of the laterals varied from 22 to 26 (maximum, 47). Although short and only fairly well rebranched, these rootlets were so very abundant that the soil was thoroughly filled with them.

Frequently, they originated in groups. On the second foot of the laterals, the branches were longer and, although still very abundant, fewer per inch. But on some roots as many as 22 rootlets per inch occurred. All were more or less thread-like and usually 0.1 to 0.5 millimeter in diameter.

They were poorly rebranched. From the widely spreading, almost horizontal roots, laterals, 3 to 12 inches in length, frequently arose. These extended in various directions, some almost directly upward. A few ended within 2 inches of the soil surface. Thus a volume of soil hitherto unoccupied was explored. But these surface-absorbing roots were relatively few and, as at the earlier examination, surface tillage would have resulted in little root damage.

Branching in the second foot of soil was at the rate of three to nine laterals per inch. Although numerous rootlets were only a few millimeters long, others extended widely. Many were fairly well rebranched. The younger portions of the roots, which occupied the third foot of soil, were furnished with only short branches. On the deeper, main vertical roots, branches were only

0.1 to 2 inches long below the 2-foot level. As shown in the drawing, their occurrence was somewhat irregular, a distribution usually explainable upon the basis of soil texture.

The roots were slightly yellowish in colour and hence the smaller ones were difficult to follow in the clayey subsoil. When they were broken, large drops of sap soon collected on the ends. This was more noticeable in the cabbage than in most other plants. The cabbage taste was characteristic of all the roots, even the still rapidly growing root ends. Although the rows were 40 inches apart, no soil space was unoccupied. In fact at this time considerable overlapping of the territory occupied by the plants in the adjacent rows and especially by plants in the same row occurred.

MATURE PLANTS

The plants were over 1 foot tall, had a spread of more than 2 feet, and possessed firm, well-matured heads. In addition to the buds or heads about 22 large green leaves per plant were present. More than a dozen dead leaves clothed the base of the stem. Thus there was presented a very large transpiring area, approximately 34 square feet, to the hot, dry, midsummer air.

Fig. 7.5 A Fully Grown Cabbage Plant Showing the Large Leaf Surface which was 34 Square Feet.

To provide sufficient water and nutrients for the tops, the root system, already so well distributed in June, had greatly extended its range and fully ramified the soil. The soil was drier in the cabbage plat than in the adjacent, unplanted area. This was very noticeable even to a depth of 5 feet. Compared to the water content in the beet plats, the surface foot was drier (often 3 to 5 per cent) at all times. At greater depths, however, there was usually less water available in the plats of beets.

A few of the main vertical roots had greatly increased in length, a maximum depth of 7.8 feet being attained. The last 8 to 12 inches of roots were quite unbranched. In fact, below 3 feet the branches were short, ranging between 0.2 and 1 inch in length. Branches 2 to 3 inches long were occasionally found. They occurred at the rate of five to nine per inch and were usually simple. These deeper roots were quite sinuous in their course. In the harder, calcareous soil of the fourth foot, the zigzag course was most pronounced.

The maximum lateral spread was scarcely greater than at the previous examination, about 3.5 feet on all sides of the plant. But within the soil volume thus delimited a wonderful root development had occurred. The abundant main roots, which ran at various oblique angles and had reached a working level of 32 inches on June 24, now attained 62 inches. Thus the volume of soil ramified by the roots was nearly doubled. The depletion of the soil moisture at this depth is clearly apparent. The roots were very abundant to the 62-inch level, being found in almost every cubic inch of soil. They were somewhat fewer just above this level and were scarce below 62 inches.

Branching in this new soil volume was profuse but scarcely so extensive as in the first 2.5 feet. Some differences in branching habit were clearly related to soil structure. The very hard, dry soil of the second foot, particularly, was not so well filled with rootlets. Here, too, the roots were much more curved and crooked, undoubtedly due to difficulty in penetrating the soil. On the best-branched portions of the root system laterals occurred at the rate of 15 to 20 per inch. They were rebranched at the rate of 10 to 12 per inch, these being branches of the fourth order.

A fair conception of the remarkable absorbing system of cabbage may be gained by a study of Fig. 7.5. It should be kept in mind that in this late stage of development branching is much more profuse than here pictured, and that a somewhat similar, although less profuse, network of roots extended to a depth of 5 feet. This filled most of the space here delimited by the widely spreading laterals. In fact the soil had been so depleted of its moisture—or perhaps this was due partly to the age of the plants—that some of the rootlets were beginning to wither and die.

SUMMARY

Copenhagen Market cabbage is characterized by a very extensive, fibrous, finely branched root system. When the taproot is injured in transplanting, one of the long laterals usually assumes the position of a taproot. It is usually no more prominent, however, than many of the other major laterals that arise in great numbers from the base of the enlarged underground part. At first nearly the entire root system consists of widely spreading branches in the surface foot of soil. Later these run obliquely downward and with other more obliquely and vertically descending laterals thoroughly occupy the deeper soil. A maximum lateral spread of 3.5 feet is attained about the time the heads are two-thirds grown, but the depth of the root system in the soil thus delimited is thereafter doubled. Mature plants have a working level of 5 feet, to which depth the soil is well ramified with a profuse network of absorbing rootlets. Thus a single plant draws upon more than 200 cubic feet of soil for water and nutrients.

EARLY FLAT DUTCH CABBAGE

The Early Flat Dutch variety was studied at Norman, Okla. The plants were spaced 2.5 feet apart in rows 3.5 feet distant. Cultivation was very shallow so as not to injure the roots.

MATURE ROOT SYSTEM

The root system of small but mature plants was quite similar in character to that just described but in extent it more nearly approached that of the June 24 examination. The strong, cord-like laterals were just as abundant and quite as well branched with delicate laterals. A maximum spread of 33 inches and a maximum depth of 46 inches were attained.

SEED CABBAGE

Plants grown the preceding summer lost their leaves during late summer, but these were replaced in the fall by new leaves, about 3 inches long, developed from the stem. The tops made practically no growth until spring and no new roots had developed. On Mar. 1 the plants were uprooted and then reset in a manner similar to that practiced in raising cabbage for seed. Conditions were favourable for root establishment, and by the end of the month a dense network of fine roots had developed from the old root-stem axis and from the broken laterals left on the plants in resetting.

Typical plants had 18 to 35 strong laterals about 1 millimeter in diameter and 11 to 14 inches long. In addition 200 to 300 roots about 0.5 millimeter thick and 5 to 6 inches long had developed. These roots either turned downward or grew in a somewhat horizontal direction through the moist, mellow soil. All of the new roots were very fine and the older portions well branched. Laterals reaching 4 inches in length occurred on the older parts at the rate of 6 to 12 per inch.

The plants had nearly finished blooming by May 8. The larger basal leaves were dead but the cauline leaves were quite green. The tops consisted of 20 to 30 main branches with long terminal racemes. The height varied from 2 to 3 feet,

The root system was extremely fine and fragile, making excavation very difficult. Moreover, branches were so numerous that it was impossible to show the entire root system in a plane. Hence, only a part of it was drawn in Fig. 7.6. This shows the dearth of roots in the 3 inches of surface soil, except directly beneath the plant; also the wide lateral spreading of the horizontal roots in the surface foot (the maximum was 36 inches); a spread of 12 to 18 inches in the second foot; and the unbranched root ends in the deeper soil. The greatest depth was 33 inches. The long, rebranched laterals intermixed with the shorter, mostly unbranched rootlets were characteristic. The latter were mostly 0.5 to 3 inches long and usually pursued a course somewhat at right angles to the roots from which they originated.

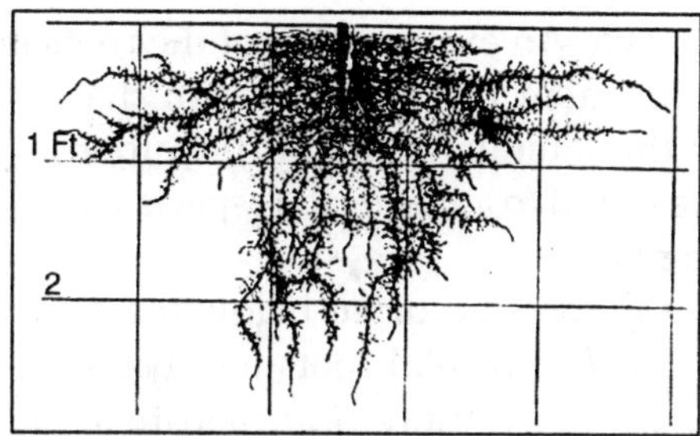

Fig. 7.6 A Portion of the Root System of a Cabbage Plant Reset the Second Year for the Production of Seed. The Old Root System has been Replaced by a new one of a Quite Different Type.

Summarizing, Early Flat Dutch cabbage has a root system similar in habit but less extensive than that of the Copenhagen Market. Plants reset for seed develop a very dense, fibrous root system which occupies a conical soil volume nearly 5 feet in diameter and 2.5 feet deep.

DEVELOPMENT OF ROOTS AND TOPS AS AFFECTED BY CULTIVATION

During 1926 experiments were conducted to determine the effects of deep and shallow cultivation upon root development of cabbage and its relation to the water and nitrate content of the soil. The plants (Copenhagen Market variety), when transplanted into the field May 3, were placed in two plats separated by an uncropped area 10 feet wide. Each plat consisted of 5 rows 40 feet long and 3 feet apart and the plants were also 3 feet distant in the rows. The field was an almost level lowland and the soil a moderately rich silt loam. Hence, the plants made a good growth notwithstanding a dry summer. In fact April, May, June, and July each had a precipitation below the normal, the total deficiency amounting to 4.4 inches. During these months only 16 showers in excess of 0.15 inch occurred. The year was also characterized by a late, cold spring. One plat was hoed 3.5 inches deep at five different periods, *viz.*, May 15 and 31, June 14 and 25, and, finally, July 11. On the same dates the other plat and the uncropped area were thoroughly scraped to a depth of ½ inch. An examination of the roots in the two plats on June 19 revealed no differences in the size or direction pursued by the laterals in the surface foot of soil. The root systems were found to agree in all respects with those excavated the preceding year.

Most of the roots occurred below the maximum depth of hoeing but some horizontal roots had been severed. In every case these were much more branched than similar, shallow, horizontal roots in the scraped plat. Moreover, from near the cut end a large number of long, rebranched laterals had extended downward. Thus the chief direct effect of deep cultivation on the root system was to promote branching of the injured roots.

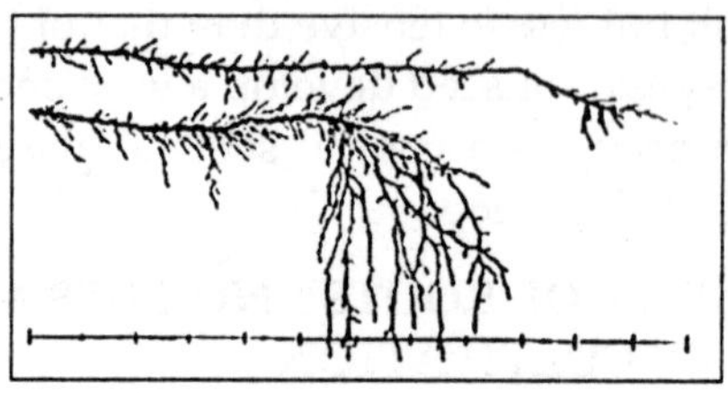

Fig. 7.7 Representative Surface Roots of Cabbage Under Two Types of Cultivation.

The upper root shows normal development in the surface soil layer. The lower one has been cut by deep cultivation. This resulted in a great increase in branching. Scale in inches.

A later examination, July 28, revealed that the roots in the scraped plat extended to within ½ inch or less of the soil surface. Under the leaves that covered the ground for a distance of 1 foot on all sides of the plant the surface soil was moist and the roots came to within 2 millimeters of the surface. In the deeply hoed plats few or no roots were found in the surface 3.5 inches of soil except near the base of the plant where branches occurred from the cut root ends. The soil was fairly moist. The center of the uncropped area was, in general, free from roots, although a few were traced two-thirds of the distance across this 10-foot strip. A study was made of the moisture content of the soil of the two plats and that of the uncropped but scraped, intervening area. Samples were taken in duplicate at the several depths at each determination. These data are given in Table. An examination of this table reveals the following facts: In both of the plats sufficient water at all times was available to promote good growth.

After the roots became widely spread between the rows, where the samples were taken, the water content in the deeply hoed plat became gradually less than that in the middle of the uncropped area. This decrease in soil moisture was in general progressive throughout the season and as the roots extended deeper it became marked even at depths below 2 feet. A decrease of 2 to 4 per cent was common in the surface foot, although it was sometimes more than twice these amounts. Moisture content of the scraped and uncropped area was also uniformly much greater at all depths than that in the scraped and cropped plat.

Determinations of nitric nitrogen were made on the several dates shown in Table. At every determination the amount was greatest in the uncropped soil; in fact, it was often twice the amount found in the cropped plats. After July 3, when the root systems were well developed, the differences were very marked. No consistent differences were found between the nitrate content of the soil in the deeply cultivated and scraped plats.

Table. Nitric Nitrogen in Parts per Million in the Several Plats

Date Inches	Depth, Area	Uncropped Hoed	Deeply Scraped	Surface
June 2	0-6	33.5	....	16.3
June 9	0-6	31.0	20.1	26.0
June 19	0-6	38.5	15.0	29.6
"	6-12	21.2	7.4	12.1
July 3	0-4	34.6	17.3	12.9
"	4-12	18.9	6.8	12.6
"	12-24	6.8	...	6.0
July 10	0-4	39.5	7.5	4.7

"	4-12	21.4	3.9	1.9
"	12-24	6.8	2.2	1.3
July 19	0-6	47.0	...	13.8
"	6-12	16.1	...	11.8
"	12-24	7.1	...	4.4

These results are in accord with those obtained at Ithaca, N.Y., where soil samples were taken at intervals of approximately two weeks during the growing season to a depth of 18 inches. They were obtained from several plats of cabbage, beets, carrots, onions, tomatoes, and celery and from fallow soil. "With the cropped areas the differences in nitrates between the cultivated and the scraped plats were not significant except with the trained tomatoes, where the cultivated soil averaged higher than the scraped soil." 159a As regards the cultivated and scraped fallow plats a slight increase in nitrate nitrogen was found in the former.

OTHER INVESTIGATIONS ON CABBAGE

In the preceding investigation in New York,... roots were found to a depth of 30 inches, even the finer roots being found in considerable numbers as deep as 24 inches. A large part of the root system, however, was found in the surface 12 inches which corresponded to the depth of the surface soil. The roots extended laterally as far as 3 feet and were about as plentiful midway between the rows as within a few inches of the plant. The roots were branched many times so that the soil was quite thoroughly filled to the depth of 6 inches, although the greatest mass was found within 3 inches of the surface. Most of the cabbage roots were quite small, but of about the same size throughout their length.

Further studies on cabbage at the same station are of interest. Twenty-five days after the plants had been transplanted into the field, they were stocky, 4 inches high, and had a total spread of 6 inches. The roots were 10 inches deep and the soil within a radius of 8 inches was well filled with fibrous roots, many of which were within an inch of the soil surface. Twenty days later, on half-grown plants the soil was well filled with fine branching roots to 15 inches depth, some extending to the 22-inch level. Lateral roots had reached the centers between the 3-foot rows and the surface soil was well filled with fine roots to a distance of 12 inches from the plant. The roots grew rapidly and when the plants were fully grown had reached a depth of 3 feet. The soil was well filled with fine roots to a depth of 30 inches. Many roots were traced to adjoining rows. The main laterals, which were of nearly the same size throughout, grew almost horizontally. The roots were extremely well branched and so filled the surface soil that cultivation could not be given without destroying large numbers of them.

RELATION OF ROOT SYSTEM TO CULTURAL PRACTICE

The very extensive and finely branched root system of cabbage, together with the extensive development of the tops, helps to explain why it is "hard on the land." Compared with beets, for example, under the same type of cultivation, Table above shows that cabbage much more thoroughly exhausts the water supply in the surface foot of soil, the place where its roots are most abundant. From May 29 to Aug. 13 an excess-moisture supply of 2 to 9 per cent in favour of the beets was ascertained. This intensive absorption by the shallower portion of the root system helps to make clear why on a good loam soil frequent shallow irrigations 8 to 10 days apart, particularly after the cabbage begins to head, result in larger yields than heavier irrigations at longer intervals. 33Indeed, it is very dependent upon a proper supply of water and suffers more from a lack of it than most garden crops.

The roots also need good aeration; in wet soils the plants turn yellow and cease growing. The roots must also supply the plant with a large amount of nutrients, especially nitrogen and potassium, to develop the large, succulent leaves. The time of application of the fertilizer influences both the time of maturity and the total crop yield. In fact, cabbage soil can hardly be made too rich if the food materials are in a well-balanced form. Thus it is easily understood why, although early varieties do well on light soils, late-maturing crops, to give high yields, must be grown on clay-loam or silt-loam soils, preferably those with considerable humus, where both water and nutrients are constantly abundant. This results in a uniform, continuous growth. If the heads, because of lack of moisture, cease growing when nearly mature and again start growth as a result of rains or irrigation, they are very likely to crack. This bursting is due to the absorption of excessive moisture by the roots. It may be prevented by pulling on the stem sufficiently to break a part of the roots or by deep cultivation.

Cultivation

The superficial position of many of the very long, horizontal roots of the cabbage and its habit of thoroughly occupying even the surface 1 or 2 inches of soil show clearly why cultivation should be shalow. Root growth is better and cultivation easier when the soil has been deeply plowed and well prepared.

Experience has shown that cabbage requires considerable room to develop properly. Consequently, plants are spaced, depending upon the variety, 1 to 2 feet apart in rows 2 to 3.5 feet distant. But after only a few weeks of growth the roots overlap between the rows and the soil is thoroughly ramified. There is no place for competing weeds and sufficient cultivation should be given to keep them out and perhaps to maintain a soil mulch until the roots rather thoroughly occupy the soil.

The plants respond to good cultivation by vigorous growth of roots and tops. They will not tolerate neglect like many other vegetable crops. Later cultivation is sure to cut many of the roots and thus decrease the absorbing area. As already indicated, this results in decreased yields. Like the soil in a field of sweet corn, it is so thoroughly occupied by roots that little water is lost by direct evaporation. In fact, the effect of a soil mulch in conserving moisture appears to have been greatly exaggerated in popular literature on gardening. Under some conditions a soil mulch conserves moisture while under others it has the opposite effect. Even where moisture is conserved by a soil mulch the advantages may be lost because of injury to the roots by cultivation when the plants are large and the root system well developed... Plants having a large and well-distributed root system respond less to cultivation, for purposes of maintaining a soil mulch, than plants with a relatively small and restricted root system. There is also less moisture conservation from cultivation with the former than with the latter type of root system. 158

Transplanting

The common practice of the successful transplanting of seedlings into the field is closely connected with root development. When it is recalled that transplanted crops such as cabbage, cauliflower, sweet potatoes, tomatoes, celery, peppers, eggplant, and Brussels sprouts comprise over half the acreage and value of all vegetables, the importance of the process may be appreciated.

Transplanting consists in lifting the plant from the medium in which its roots are established and in replanting in a different location. It is a violent operation because the younger roots with their root hairs are, as a rule, sacrificed in the process of lifting. This is just the part of the root system most active in absorption. Taking up plants for transplanting results not only in breaking many of the roots but also especially in injury to the taproot. As a consequence many new roots are formed. These do not grow so long as the original ones. They form a more compact root mass about the base of the plant. Hence, the root system is less disturbed when the plant is finally transplanted into the field. Thus, although the root system of transplanted plants may be less extensive than that of undisturbed ones, upon removal to the field the transplanted plant carries more roots with it and consequently more readily reestablishes itself.

Recent investigations throw much light upon the results obtained by different methods of transplanting vegetable crops. Like many old-world practices brought to America, such as pruning and suckering, excessive cultivation, and too heavy fertilizing, methods of transplanting have heretofore been accepted without adequate investigation. Because of the excellent tilth and fertility of virgin soils the growing of vegetable crops has been generally prosperous in spite of these practices some of which, however, have been gradually eliminated either as being unnecessary or injurious.

Many of the data on transplanting have been abstracted from the excellent paper: Loomis, W. E., "Studies in the Transplanting of Vegetable Plants."

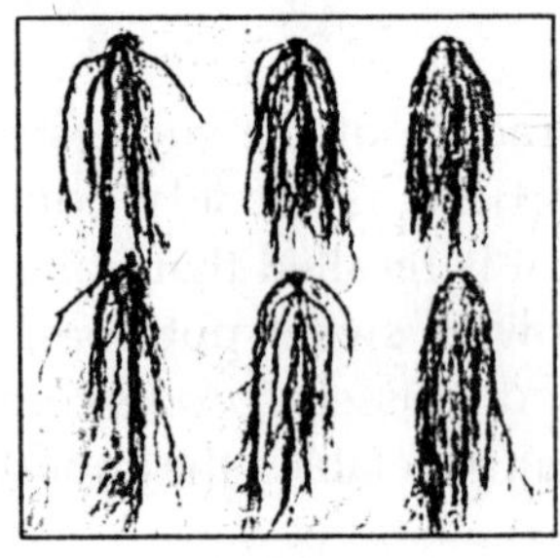

Fig. 7.8 Effects of Transplanting on the Root System of Cauliflower. Left, not Transplanted; Center

One of the chief arguments nearly always advanced in favour of transplanting is that the branching of the root system is increased as a result of the root pruning. That the roots of transplanted plants are more branched than those not transplanted is shown in Fig 7.9. The tendency of transplanted cabbage to retain balls of earth with their roots as well as the development of the root system itself are shown in Fig. Thus it is possible to transfer a good ball of earth with twice transplanted cabbages, for example, while the roots of plants not transplanted are practically bare. Although transplanting was formerly considered beneficial in itself, it is gradually coming to be recognized as an expedient not directly promoting the development of the plant. Although strongly recommended by the older gardeners and many of the earlier writers on the subject of vegetable growing, there seems at present to be considerable doubt in the minds of growers concerning the advisability of giving more transplantings than are required in the most economical production of the crop. That is, the tendency is away from transplanting as a cultural practice towards transplanting as a matter of economics.

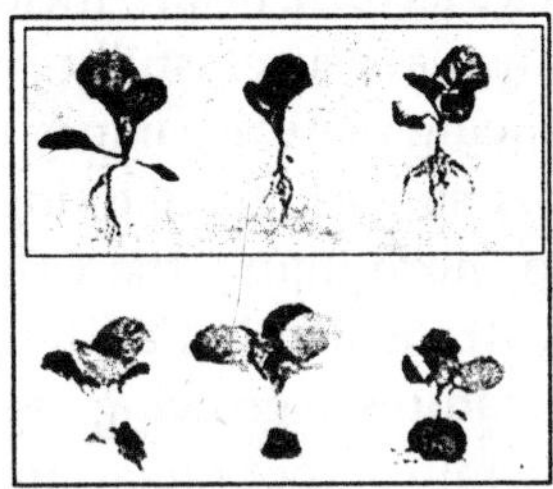

Fig. 7.9 Effects of Transplanting upon the Root System of Cabbage and the Tendency to Retain Earth about their Roots. Left, not Transplanted; Center, once Transplanted; Right, Twice Transplanted.

Certain crops, as early lettuce and cabbage or tomatoes, are transplanted in order to grow them out of the normal season for a given locality. Tomatoes, sweet potatoes, and similar vegetables may also be grown in higher latitudes if the plants are started with artificial heat. The use of artificial protection usually requires transplanting, because it is most convenient to have the plants concentrated into a small area while the protection is being given. The same is true when special care in cultivation, watering, protection from insects, and so forth are required by the seedling plants. A third factor, which accounts for most of the transplanting in greenhouses and some of that done in intensive cultivation, is the saving of space and of expensive seeds. For these reasons transplanting will be practiced as long as profitable crops can be produced by the method. On the other hand, it is an expensive operation and not to be performed unnecessarily.

Experiments have shown that the general effect of transplanting is to retard development. The growth of young plants may be arrested without serious injury but the effect of equivalent checking increases with the maturity of the plant. The degree of retardation varies with the kind of plant, its age, and the conditions of transplanting. Cabbage is one of a group of plants, in which are also included tomatoes, lettuce, and beets, that easily survives transplanting. Peppers, onions, celery, and carrots are transplanted with more difficulty and a third group consisting of such species as corn, beans, melons, and cucumbers are very difficult to transplant successfully except at a very early age. In fact with advancing maturity injury from transplanting increases in all cases. A single transplanting at a later stage of development may do more injury than two or three earlier transplantings. Hence, transplanting at the proper time is one of the most important features in growing vegetable crops. There is a rapid decline in the rate of root replacement with increasing age.

The immediate effect of transplanting is to slow down or stop the growth of the plant for a period which seems to vary directly with the amount and duration of the reduction of the water supply. The rate of new root formation is the most important consideration in the reestablishment of transplanted vegetable crops. When a large proportion of the root system is retained and adequate moisture supplied, there may be very little harmful effect from transplanting.

Recovery from transplanting is affected by a large number of environmental and internal conditions. Among these are the amount of suberization of the older roots, the proportion of the root system normally retained in transplanting, the rate of new root formation, and adaptation of the tops (by hardening, etc.) to prevent water loss or increase resistance to death by sudden wilting. In addition to these are the environmental factors of soil moisture, humidity, temperature, wind, etc. But all appear to be based in their final effect upon a change in water supply of the plant. Experiments have shown that the most important factor involved in resistance to transplanting or recovery is the root. Cabbage and other easily transplanted plants, as compared with corn, melons, etc., retain a relatively larger proportion of their root system when transplanted (a fact due in part to their network of fine branches); the retained roots are much less suberized and consequently more efficient absorbers; and the rate of new root formation is much greater.

The practice of growing plants in pots, small boxes, cans, or plant bands is advantageous inasmuch as the root system is little disturbed in setting the plants into the field, since practically all of the roots remain in the ball or block of soil. Likewise, they may be shifted from one receptacle to another of larger size with little or no injury to the roots.

In transplanting cabbage the leaf area is often considerably reduced so that transpiration will not be too great. This is accomplished by gathering the leaves of the plants together and shearing the upper portion, care being taken not to injure the buds. The transplanting of smaller plants, however, would seem more advisable. Although the ratio of roots to tops may fluctuate widely for a short period after transplanting, it tends to come to an equilibrium which is dependent upon the balance between growth and food supply in the plant, *i.e.,* the supply of water and soil nutrients to the tops and elaborated food to the roots. Further studies of these subjects are needed.

The advantage of using only the strong, stocky plants has been fully demonstrated. As a result of an experiment in Pennsylvania extending through a period of 3 years, it was found that small plants for the late crop gave a yield of only 12.7 tons per acre; plants of medium size 17.7 tons; and large plants 21.0 tons per acre. Enkhuizen Glory and Danish Ballhead varieties were used and the plants were graded according to size at the time they were planted in the field.

Experiments have shown, contrary to general belief among gardeners and others, that transplanting in itself neither increases the yield nor hastens maturity. These conditions result when decreased competition among the plants is brought about by giving each a greater space for growth. Extensive experiments in greenhouse and garden have shown that when transplanted plants are allowed no more space than those not transplanted, the effect of transplanting was, in every instance, a reduction in the yield. For example, when cabbage was grown in the field and alternate plants in the row taken up and reset in the same place, the yield of 19 heads was 119.3 pounds as compared to 132.6 pounds from a similar number of plants not transplanted.

An experiment conducted in the greenhouse gave similar results:

- 8 plants not transplanted weighed 4,214.0 grams
- 8 plants once transplanted weighed 2,993.5 grams
- 8 plants twice transplanted weighed 2,241.7 grams

Thus plants once transplanted yielded 28.9 per cent less and those twice transplanted 46.8 per cent less than those whose root systems were not disturbed. Similar results were obtained with two varieties of cabbage and with cauliflower, tomatoes, and lettuce.

RELATION OF ROOTS TO COMPETITION AND DISEASE

Competition among cabbage plants must be entirely below ground and in connection with the root systems, for the widely spaced plants will not shade each other considerably. Hence, attention in spacing should be fixed upon root habit in the various types of soil in connection with the size of head desired and methods of cultivation.

Certain diseases are closely related to root development. Clubroot is due to a parasitic slime mold (*Plasmodiophora brassicae*) which attacks not only cabbage but plants related to it such as cauliflower, kohl-rabi, turnip, rutabaga, etc. The roots become enlarged and malformed and fail to function normally. As a result the plant is subject to wilting during periods of high transpiration. The plants do not thrive and yields are greatly reduced, if indeed the plants do not die. Root knot is a somewhat similar malady, common in the South, caused by a parasitic eelworm.

It may affect all kinds of vegetable crops. In cabbage yellows, a soil-inhabiting fungus gains entrance through the root hairs, pushes back through the cortical tissue, and grows throughout the vascular system. These invasions of root and stem result in diminution of water and supplies of food materials from the soil, which, in turn, give rise to stunted plants. The host may be killed in the seedling stage or wilt and die at any time during its growth. The leaves have a pale, lifeless, yellow colour. Sometimes only one side of the root system is seriously attacked. Then the opposite side of the plant grows more rapidly and brings about a curving of stem and leaves. Both root and stem are greatly dwarfed. The majority of the diseased plants continue a sickly existence for a month or more and then succumb.

Investigations have shown that disease resistance and predisposition to disease may largely depend upon environmental conditions under which the plant is developing. By selecting plants whose roots are resistant to the fungus attack or ensuing injury, cabbage yellows has been brought under control. These facts suggest another reason for a thorough understanding of not only the morphological relation but also the structural changes and chemical composition of roots.

CAULIFLOWER

Cauliflower (*Brassica oleracea botrytis*) is a cole crop very closely related to the cabbage. It is a variety of the same species, both having originated, probably as mutants, from the same wildcabbage ancestor. The cabbage head is a bud, but that of cauliflower consists of fleshy peduncles, pedicels, and other flower parts subtended by a number of cabbage-like leaves. Cauliflower does not thrive in hot, dry weather and consequently is grown either as an early or late crop. It is not as hardy, can stand less heat, and is much more exacting as to climate than cabbage. Hence, on a large scale it is grown only in a relatively few restricted areas in the United States.

Seedlings of the Early Snowball variety were transplanted into the experimental field at Norman, Okla., Apr. 19. The plants, which were in the third- to fourth-leaf stage of development, were set in rows 2.5 feet apart and 3 feet distant.

EARLY DEVELOPMENT

The first root excavations were made May 10. The plants had made a vigorous growth. There were 6 to 8 leaves, 3 to 4 inches long, per plant. The taproot, as usual, had been broken in transplanting. The root system consisted of a large number of delicate white roots less than 1 millimeter in diameter. As shown in Fig. 7.10, many of these ran horizontally just beneath the soil surface, some to a distance of 15 inches. Others ran outward and downward and still others descended almost vertically. The greatest depth was 15 inches. Laterals 2 to 4 inches long had grown from the older parts of the largest roots at the rate of 8 to 16 per inch. Some of these gave rise to short sublaterals. A similar branching rate was evident throughout the plant, but tertiary branches had not yet been formed on most of the rootlets. The thorough occupancy of the surface soil was possible because of very shallow cultivation.

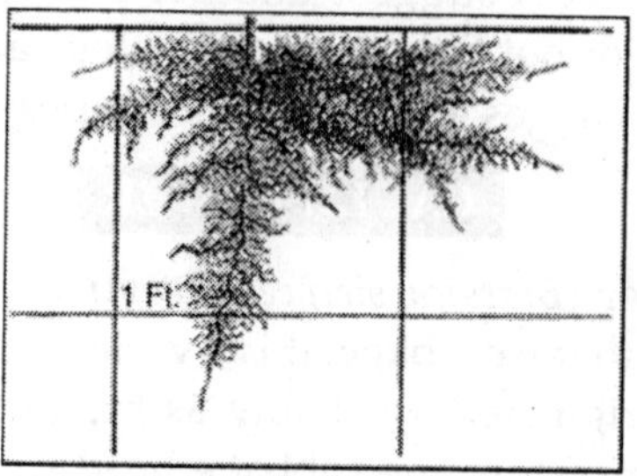

Fig. 7.10 Root System of Early Snowball Cauliflower 3 Weeks After Transplanting into the Garden. Note the Fibrous Roots Near the Soil Surface.

HALF-GROWN PLANTS

The plants grew vigorously and by June 15 the stems were over 0.5 inch in diameter and the plants 1 foot tall. Leaves were abundant but "heading" had not begun,

Typical plants had 6 to 10 main roots which were 1 to 3 millimeters in diameter. Due to frequent branching they tapered rapidly so that 18 inches from their origin they seldom exceeded 1 millimeter in thickness. Some extended widely in the surface foot of soil, others penetrated almost vertically downward, and still others pursued an intermediate course. Hence, the general shape of the root system was that of an inverted cone with a diameter of about 2 feet at the base and a depth of 2.5 feet (Fig. 7.11). All of the roots were profusely branched with both long and short laterals. The larger ones were usually 5 to 9 inches in length but a few 15 inches long occurred. Near the base of the plant and especially on the larger branches rootlets, often 4 to 10, grew in clusters. These root clusters were very numerous, frequently 12 to 15 occurring per inch of main root. The rate of branching elsewhere was about 16 laterals per inch. Thus the soil volume was exceedingly well ramified by an extensive network of delicate, white rootlets.

Fig. 7.11 Cauliflower 8 Weeks after Transplanting into the Garden.

MATURING PLANTS

When the cauliflower was suitable for table use, July 19, further examination of the root system was made. The plants were 2 feet high and had a spread of tops of 2.5 feet.

A typical root system had 17 strong branches 3 to 10 millimeters in diameter. Few of them gave rise to major laterals until they had reached a distance of 6 to 12 inches from the base of the plant. Thereafter they branched repeatedly as shown in Fig. 7.11. The general shape of the root system had not been greatly changed except that a much larger soil volume in the second foot was now occupied. In this soil layer the lateral spread had been extended from about 8 inches to 2 feet.

The third foot of soil was also moderately well ramified, the deepest root, which had apparently replaced the taproot, reached the 54-inch level. The soil was also well occupied by rootlets near the surface. A lateral spread of 2.5 feet was found. Branching was even more profuse throughout the whole volume of moist, mellow soil than at the preceding examination. In fact, it is quite impossible to show the complete degree of branching in the most carefully executed drawing.

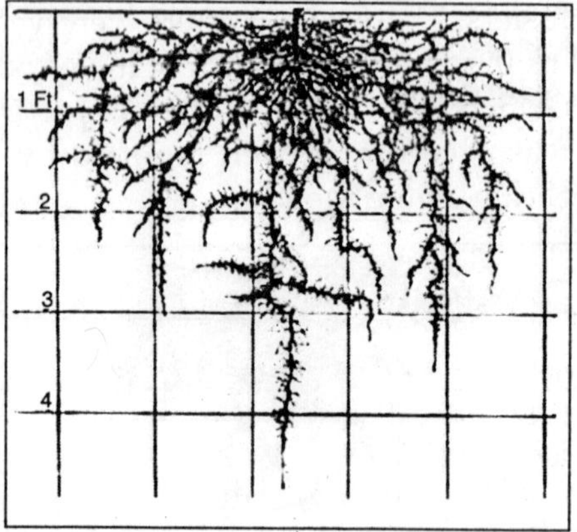

Fig. 7.12 One-half of the Root System of Cauliflower.

SUMMARY

Cauliflower, like cabbage, is characterized by a taproot which is usually injured in transplanting. Numerous, large, exceedingly well-rebranched laterals and a very profuse network of smaller ones occupy a large soil volume. Early in the development of cauliflower the soil from the surface to a depth of 8 to 12 inches is filled with fine roots throughout a zone with a radius of 1.5 to 2 feet. The 2 feet of soil beneath the plant is likewise well ramified.

The widely spreading, main laterals finally turn downward. With their profuse network of branches, aided by those of more obliquely penetrating laterals, they thoroughly ramify the first 2 to 3 feet of soil throughout a territory extending 2 to 2.5 feet on all sides of the plant. The deeper soil to 4 feet is less fully occupied by vertically descending roots or branches. As a whole the root system even more thoroughly fills the soil than does that of cabbage.

OTHER INVESTIGATIONS ON CAULIFLOWER

Few root studies have been made on cauliflower. It has been

examined at Geneva, N. Y., and found to be a deeply rooting plant. Many roots were traced downward to a depth of 2.5 feet and some reached a depth of 3 feet. The horizontal root extent was 2.5 feet on all sides of the plant. Fibrous roots were less numerous in the upper layers of the soil than on tomato plants which were examined at the same time.

ROOT HABITS IN RELATION TO CULTURAL PRACTICE

The root systems of cauliflower and cabbage have many similarities and the method of planting and growing cauliflower is very much the same as for cabbage. Climate is a more important factor than soil; the vigorous and extensive root system develops well in nearly all kinds of soil. Growth is promoted by thorough soil preparation and the liberal application of manure or other fertilizers. A constant supply of moisture for the roots should be maintained so that growth of the plants is never checked. This may be promoted by frequent tillage. Cultivation, as for cabbage, should be shallow. Because of the thorough occupancy of the surface-soil layers by the maturing root system, it would seem that, as in the case of cabbage, late cultivation might do more harm than good. All of the soil between rows spaced 2.5 to 3.5 feet apart is thoroughly occupied.

8

Rabi, Turnip and Rutabaga

KOHL-RABI

Kohl-rabi (*Brassica caulorapa*) is a low, stout biennial, closely related to cabbage. Unlike cabbage and cauliflower it does not form a "head" but is cultivated for the fleshy stem which is produced the first season. This originates above the cotyledons, is short, 3 or more inches in diameter, and grows just above the ground. This vegetable is grown chiefly as an early spring crop, since it does not thrive during the heat of summer.

Seed of the Early White Vienna variety was sown Apr. 27, in rows 18 inches apart. Later the seedlings were thinned to 12 inches distant in the row.

EARLY DEVELOPMENT

The plants were characterized by strong taproots and numerous, nearly horizontal, widely spreading laterals in the surface 8 inches of soil (Fig. 8.1). Several of the main roots ended at depths of 2.5 to 3 feet. Just below the soil surface the slightly thickened taproots were clothed with very numerous, short, unbranched laterals which formed a dense network. Usually about 20 larger roots also arose in the first 10 inches of soil. Some were only 6 inches long, many extended outward 1 to 1.5 feet, and a few of the longest (usually about 6 inches deep) spread laterally 3 feet. These larger roots were furnished with an abundance of short, mostly unbranched laterals. They averaged less than 1 inch in length and occurred at the rate of about 6 per inch of main root. Below 8 inches, branches were fewer, especially between 8 and 16 inches, and mostly less than 1 inch long, although a few exceeded 5 inches and were quite well rebranched.

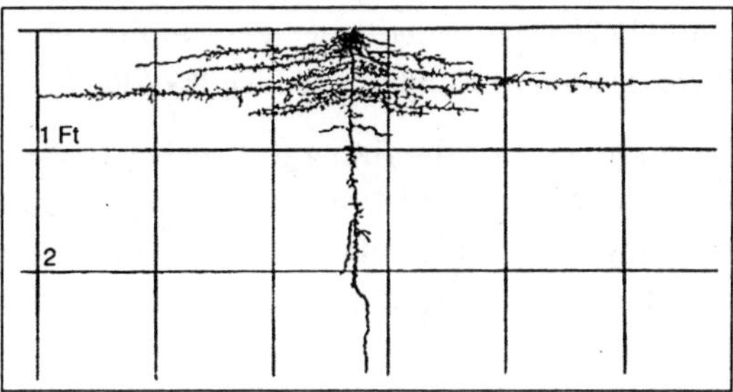

Fig. 8.1 Early White Vienna Kohl-rabi about 6 Weeks Old.

LATER DEVELOPMENT

A second examination was made 20 days later, June 30. The thickened portion of the stems had a diameter of approximately 3 inches. The plants were 1 foot tall and each had about 18 large leaves. The large transpiring surface is indicated by the fact that the leaf blades averaged 7 inches in length and 6 inches in width.

During the 20-day interval the roots had grown considerably. Moreover, the number was greatly increased. Numerous counts showed that a typical plant had 81 nearly horizontal laterals in the surface inch of soil, 70 in the second inch, and 10 in the third. Between 4 and 9 inches they occurred at the rate of about 6 per inch. Although many of these did not exceed 3 to 6 inches in length, fully 20 per cent extended rather horizontally to greater distances, the longest having a total spread of 40 inches (Fig. 8.2). The larger laterals had diameters of 2 to 3 millimeters.

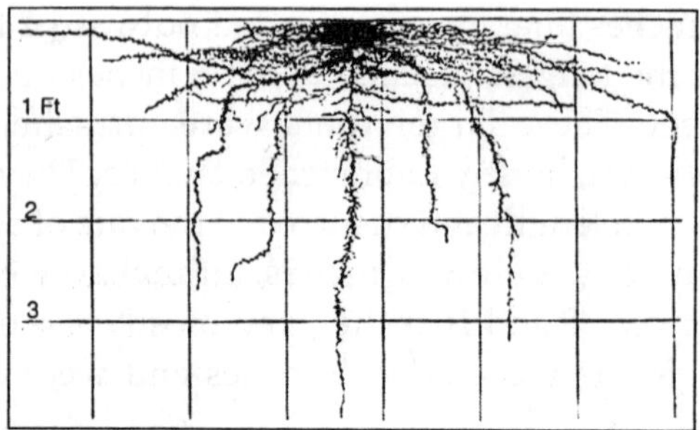

Fig. 8.2 Kohl-rabi 20 Days Older than Shown in Fig. 38. Note the Increase Both in Number and Length of Roots.

Near the plant, *i.e.*, for the first 4 to 8 inches of their course, the roots were extremely well branched, rootlets quite filling the soil. Branching was at the rate of 12 to 20 laterals per inch. They were 1 to 2 inches long and gave the main branches almost a woolly appearance.

The major portion of the laterals was furnished with branches 0.2 to 1 inch in length at the rate of 4 to 8 per inch. Branches of the third order, although not long, were very common. The main laterals also had large branches which ran in various directions, very often towards the soil surface (Fig. 8.3). These were rebranched in a manner similar to that of the large roots. Thus the surface 10 inches of soil to a distance of at least 2 feet on all sides of the plant were well occupied by roots. Near the plant the roots occurred in dense masses forming a thread-like network. That the horizontal branches were rapidly growing was shown by their long, thick, unbranched ends.

Below the 10-inch level only a few roots were found. Aside from the taproot these consisted of three to five of the obliquely descending laterals or horizontal ones which had spread 9 to 15 inches and then turned downward. The smooth, unbranched ends of these were usually found between the second and thirdfoot level. Branching was abundant but not profuse. Between the 10- and 18-inch level the taproots usually had about five branches per inch. They varied considerably in their direction of growth but were seldom over 6 inches in length.

At greater depths the roots became shorter (3 inches or less) but were moderately well rebranched to near the 3-foot level. Some of the taproots reached a depth of 4 feet. A study of the root system clearly showed that the bulk of absorption was still taking place in the surface soil. Tillage, except of the most superficial kind, would have been distinctly harmful to the roots.

MATURE PLANTS

At the time of the final examination, Aug. 13, the plants had scarcely increased in size or number of leaves. The leaves were somewhat larger, however (about 10 inches long by 7 inches wide) and the fleshy portions of the stem now had a diameter of 4.5 to 5 inches.

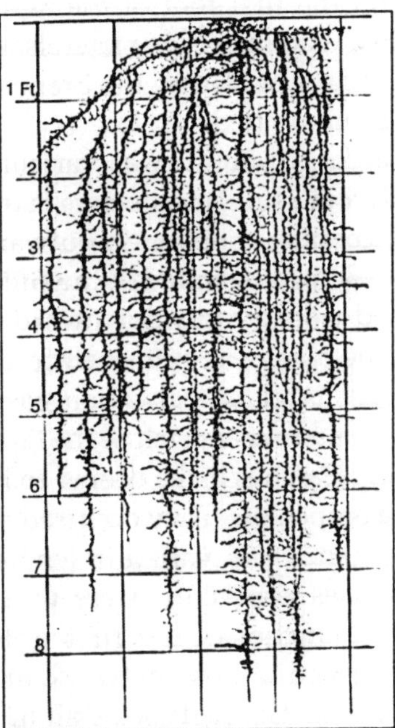

Fig. 8.3 Root System of a Nearly Mature Kohl-rabi. Branches on the Fine Lateral Roots arising Directly from the Taproot are not Shown.

The underground part had made a remarkable growth, especially in depth. The lateral spread of any of the plants examined did not exceed that of the June examination. The fleshy tap-roots frequently measured 1.7 inches in diameter near the soil surface but tapered to about half this diameter or less at the 6-inch level. At greater depths they scarcely exceeded some of the larger laterals which were 6 to 10 millimeters in thickness. They pursued a vertically downward course to a maximum depth of 7 to 8.5 feet.

As at the earlier examinations, most of the larger branches arose from the first 10 inches of the taproot. In general, the roots here were of two sizes. There were small ones, usually less than 1 millimeter in diameter and only 1 to 6 inches long, and others distinctly larger and very extensive. They originated from two sides of the taproot. Frequently as many as 140 of the smaller laterals occurred on the first foot of the taproot. They were profusely rebranched with rather long laterals which were again rebranched. For the sake of clarity, the branches on these fine lateral roots have been omitted.

The number of large laterals was variable, 6 to 10 being usual. They ranged from 2 to 10 millimeters in diameter. Usually they ran rather horizontally or obliquely outward 1 foot or more and then, turning downward, ended in the fifth to seventh foot of soil. Others turned more directly downward so that the entire soil volume below the plant was quite well occupied. As shown in the drawing, these large laterals often gave rise to strong branches. All these main roots as well as the taproot, which also gave rise to a few large laterals in the deeper soil, were profusely furnished with fine branchlets. They occurred at the rate of 5 to 10 or even more per inch and varied in length from 0. 1 to 8 inches. The longer ones, especially, were well rebranched. In general, the small branches of the first order were rather horizontal. They formed a close network in the soil to the working depth of 7 feet. The roots were all in good condition and growth was still taking place.

SUMMARY

Young plants of kohl-rabi are characterized by a well-developed taproot and very numerous, shallow, but widely spreading, horizontal branches. In this respect it is very similar to cabbage. By July 1 the surface foot of soil 2 feet on all sides of the plant is ramified by a dense network of rootlets. A few obliquely descending rootlets and the taproot extend into the third or fourth foot of soil. In mature plants the taproot below a depth of 1 foot scarcely exceeds in importance the 6 to 10 major branches or even some of their laterals which, after running outward, parallel its course. Both taproot and branches reach depths of 5 to over 8 feet. All are profusely furnished with fine, rebranched laterals, often of considerable length. From the dense network of short, surface branches to the working level at 7 feet, the soil is thoroughly ramified.

OTHER INVESTIGATIONS ON KOHL-RABI

Few investigations have been made on the root system of this plant. At Geneva, N. Y., the taproot of the Early Purple Vienna small variety was traced to a depth of more than 2 feet, having been followed through 14 inches of very compact clay. "Owing to the delicacy of the root we were unable to reach the end." The horizontal roots extended 2 feet on all sides of the plant. As with cabbage and cauliflower, it was found that the fibrous roots were most numerous in the upper 8 inches of the soil. Investigations in Germany have shown that the main mass of the roots of kohl-rabi are in the upper soil layers, but in the varieties of cabbage examined the soil was thoroughly ramified to a depth of 39 inches. Lateral spread varied from 3.3 to 4 feet. These observations were made in the main by means of plants grown in containers, although also aided by observations from plants growing in the open, the root systems of which were carefully washed from the soil. Although it seems clear that differences in subsoil conditions greatly affect the depth of root penetration, it is equally certain that the roots of this plant may penetrate deeply.

TURNIP

The turnip (*Brassica rapa*) is a vegetable found in many home gardens throughout the United States. Since it is a very hardy plant of rapid development but does not thrive in hot weather, it is usually grown either early in the spring or late in the fall in the North and as a winter crop in the South. Although the large rosette of coarse, roughish leaves is sometimes used for "greens," turnips are usually grown for their enlarged fleshy roots. The upper portion of this enlargement to which the leaves are attached is morphologically a stem. During the second year of growth, or late in the fall of the same growing season, if planted in the spring, it produces a large, branched flower stalk 1 to 3 feet in height. Thus the turnip is either an annual or a biennial.

The Purple Top Globe turnip was planted June 16, 1 month too early for its best development, in rows 16 inches apart. After the plants were well established, they were thinned to 6 inches distant in the row. By July 10 they were 5 inches tall and had 5 to 7 leaves each. The five largest were 3 to 4 inches long and 2 to 3 inches wide, thus presenting a rather large transpiring and photosynthetic surface.

EARLY DEVELOPMENT

The turnip is characterized by a strong taproot which was at this time 5 millimeters thick at the ground line but soon tapered to a diameter of 1 millimeter or less. It pursued a somewhat tortuous downward course, making gentle curves through an amplitude of 1 to 3 inches. Depths of 30 to 32 inches were attained by the taproots, the last 3 to 4 inches of which were unbranched. The first 2 inches of taproot were also free from branches; there were 9 to 10 laterals per inch on the next 4 inches on typical plants. Below this they varied from 6 to 12 per inch to near the root ends. Thus a total of 200 to 225 branches occurred on a single plant (Fig. 8.4).

It may be noted that most of the laterals in the surface foot spread somewhat horizontally. At greater depths some ran more or less horizontally to only 1 to 3 inches and then turned downward. The longest branches and greatest lateral spread (about 14 inches) occurred in the first foot. Practically all of the laterals were clothed with rootlets less than 0.1 to 0.3 inch long at the rate of two to five per inch.

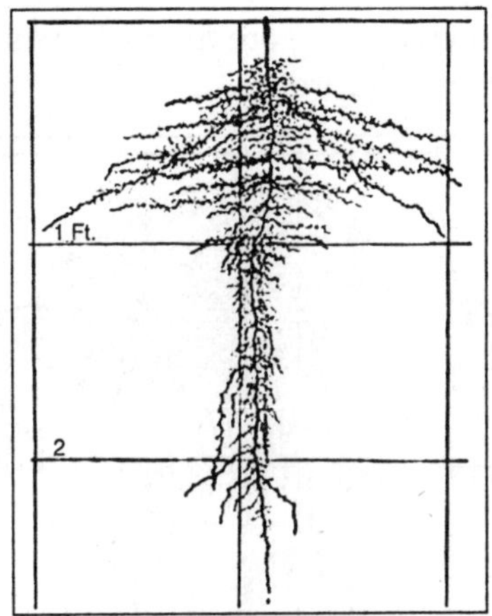

Fig.8.4 Taproot System of Purple Top Globe Turnip in its Early Development.

On the largest laterals they were longer and often much more abundant. Where they exceeded 1 inch in length, branches of the third order were found at the rate of three to five per inch. Thus an extensive absorbing surface was already developed.

MIDSUMMER GROWTH

The tops were 1 foot high and each plant had about a dozen leaves. At least eight of these averaged 8 by 5 inches in length and width, respectively, although some leaf blades were nearly 1 foot long.

Correlating with the growth of tops, the roots too had made a marked growth. A maximum lateral spread of 2 feet had been attained and many of the taproots reached the 4-foot level. Thus the root system nearly doubled in depth and lateral spread (Fig. 8.5). The taproots were now about 0.8 inch thick but usually tapered to 2 millimeters or less at the 6-inch level where they were scarcely larger than many of the major laterals. As a result of recent showers, thread-like but very profusely rebranched laterals filled the surface soil. Although only 1 to 1.5 inches long, they were rebranched to the third order. The number of lateral roots had greatly increased, more than 100 usually being found on the first 12 inches of taproot.

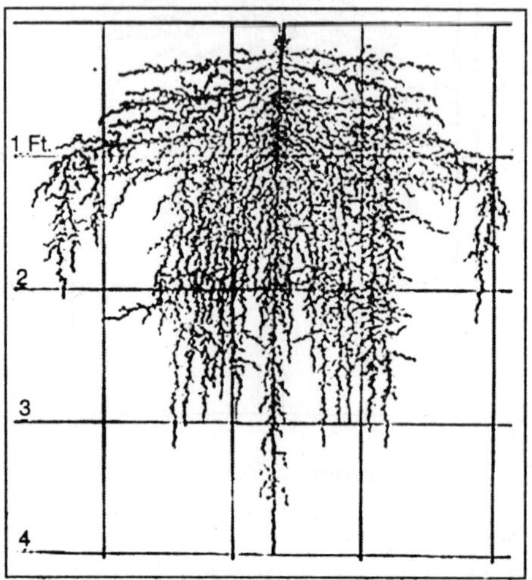

Fig. 8.5 Root Development of Turnip on July 27.

Frequently 12 to 18 major laterals originated at depths of 3 to 10 inches. As before, these ran outward in a generally horizontal direction, some ending at distances of 9 to 18 inches from their origin. But perhaps more frequently they extended outward and then turned rather vertically downward, frequently reaching the 3-foot level. Still others, especially those originating between 6 and 14 inches, ran obliquely outward and downward thus occupying the soil nearer to the taproot.

Below 14 inches few long branches arose from the taproot, but it was clothed with an abundance of short roots (0.5 to 4 inches long) which usually occurred at the rate of about 10 per inch. Branches on all of the laterals were abundant; 8 to 12 per inch were common and sometimes there were as many as 15 to 20. Many were hairlike, others were larger and 2 to 18 inches long. They extended in all directions and gave rise to sublaterals which were often 3 to 5 inches long and themselves rebranched. In fact the network of rootlets was frequently so dense as to form brushlike mats which thoroughly ramified the soil for water and nutrients.The difficulty of the roots in penetrating the stiff subsoil was shown by the characteristic and abrupt kinks and turns and often zigzag course of the taproots and by both large and small laterals.

MATURE PLANTS

A final investigation was made Oct. 5. The plants were practically mature. They had a height of 8 to 10 inches, a spread of tops of 15 inches, and 10 to 12 leaves each. The leaves averaged 13 inches long and 5 inches in width, thus presenting a very large area.

The fleshy portions of the roots were not well developed, the greatest diameter being only 3 inches and the long axis about 4 inches. Rootlets arose from the lower half of the "turnip" in two opposite rows. The larger rootlets were never very abundant. Only 3 or 4 per plant exceeded 1 millimeter in diameter, and these were only 6 to 8 inches long but densely branched to the third order, The rest constituted little tufts of hairlike roots, often 5 to 10 arising in a cluster at the rate of 5 to 10 clusters per inch. These were so densely branched and rebranched as to constitute a root network in the soil. Below the enlarged fleshy part, the taproot tapered to only 0.5 inch in diameter at the 12-inch soil level.

A maximum depth of 5.5 feet was ascertained for several plants. Not infrequently the taproots divided in the hard subsoil at depths of 3 to 4 feet into two to four strong branches. These often spread a few inches and then pursued more or less parallel but tortuous downward courses. The newer and deeper portions of the taproot, which had grown since the July examination, were very profusely branched.

Briefly, the chief developments of the root system since the last examination were as follows: The numerous laterals from the shallower portion of the greatly enlarged taproot and its more superficial, horizontal branches quite filled the surface soil. The lateral spread had reached a distance of 2 to 2.5 feet on all sides of the plants. The marked growth of the taproot into the deeper soil had been accompanied by a similar growth of many of the vertically descending, major branches. Thus the soil all about the plant was thoroughly occupied by a dense network of roots to the working level at about 5 feet. The degree of branching was even greater than before; even the deeper soil (below 3 feet) was literally filled with cobwebby networks of delicate rootlets, the ultimate branchlets of which were exceedingly numerous. Thus the turnip had developed a wonderfully efficient absorbing system.

ROOT DEVELOPMENT THE SECOND SEASON

The root development of the Purple Top Globe turnip at the time of flowering was studied at Norman, Okla. Plants were grown in rows 3.5 feet apart. The seed was planted about midsummer and root studies made May 1 of the following season. Early in March the plants were thinned to 2 feet apart in the rows. Earlier investigations had shown that new root growth had started and was well under way by Mar. 9, synchronously with the new growth of tops. This consisted of an abundance of young rootlets 0.2 to 1 inch in length originating from the taproot and the older portions of its major branches, Moreover, it was found that practically the entire root system had survived the winter, the soil moisture being very favourable and soil temperatures not extremely low.

On May 1 the plants had just completed flowering. The death of most of the basal leaves and the smaller size of the cauline ones had greatly reduced the transpiring area. The latter, however, were quite abundant on the 10 to 17 strong branches per plant. A rapid deterioration and death of the plants after seeding was observed.

The strong taproots pursued a very tortuous downward course to depths of nearly 4 feet. In the first foot of moist, fertile soil 58 branches 0.5 to 1.5 millimeters in diameter originated. These ran horizontally outward 20 to 41 inches and ended in the surface 12 to 16 inches of soil without turning downward. Six other branches were traced outward and downward, their courses being somewhat parallel to that of the taproot. In addition, there were exceedingly numerous small branches arising from the taproot and the fleshy portions of the major laterals. These varied from 800 to over 2,000 in number depending upon the size of the root and the number of larger branches. They were fewer where larger branches were-more numerous. Approximately one-third of these were 9 to 18 inches long, the others were shorter. Most of them had appeared since the March examination when only 200 to 500 roots of this type were found. Although some were simple, they wore usually well branched.

Branching on the major laterals was profuse. A rate of 15 to 24 thread-like branches per inch was not uncommon. Some of these were 9 inches long and profusely rebranched. In places as many as 86 rootlets 1 to 3 inches long were found on a single inch of main root. In general, the laterals pursued a path at right angles to the root from which they arose.

In the second foot of soil a total of 15 larger laterals 8 to 24 inches in length occurred. There were in addition 12 to 20 smaller roots on each inch of the taproot. The larger ones, in the first foot of their course, gave rise to 1 or 2 major branches '12 to 15 inches long, and 6 to 15 smaller ones 0.5 to 3 inches long. Branching in the third foot of soil was only slightly less pronounced, several strong laterals originating at this level. Even in the fourth foot well-branched and rather numerous root termini were found. The larger, longer laterals and their greater abundance is believed to be due to the sandier nature of the soil here than at Lincoln.

As a whole the roots of the turnip form an extremely intricate network and thoroughly ramify the soil, rather completely exhausting it of its supply of water. The older roots were yellowish in colour, the younger ones and the root ends were white.

SUMMARY

The turnip has a pronounced taproot which elongates at the rate of 1.2 inches per day during the first few weeks of growth. In the surface foot of soil most of the branches, few of which are near the soil surface, extend horizontally but those originating at greater depths very early show a tendency to turn downward. Plants only 3 weeks old have taproots over 2 feet in length from which an extensive and profusely branched absorbing system is soon developed. Forty-day-old plants extend their taproots to the 4-foot level. Lateral spread in the surface foot reaches 2 feet, but the surface 4 inches of soil is poorly occupied.

Thus in a period of 17 days the depth and lateral spread of roots have been nearly doubled. By a turning downward of the horizontal roots and a marked growth of more oblique ones, all of which originate in the surface foot, the second and third foot of soil are also exceedingly well occupied. Short branches form a dense absorbing network along the path of the taproot. Mature plants are rooted 5.5 feet deep and have a lateral root spread of 2 to 2.5 feet. By vigorous growth of the numerous vertically descending major branches, the working level has been extended to 5 feet. Thus a soil volume of 100 cubic feet is thoroughly ramified (but in the surface 3 inches near the plant only) with an extremely extensive and delicate system of ultimate, absorbing rootlets.

In more sandy soil and during a second year of growth the same variety reaches a depth of about 4 feet, develops an extensive, absorbing network in the surface soil, and has a more horizontal spread of branches throughout. Moreover, the secondary branches are much longer and ultimate branches far more numerous.

OTHER INVESTIGATIONS ON TURNIP

The Following Observations were Made at Geneva

The roots of a plant of the Purple Top Globe turnip were washed out Oct. 29. The root weighed 3 pounds and the foliage was very vigorous, but the scanty root development was a matter of surprise. The deepest root appeared to extend no more than 18 inches and the longest horizontal roots reached no further. The taproot tapered rapidly from the bottom of the bulb for a distance of 6 inches, where it divided into two branches, each about 1/16 inch in diameter. Only 13 branches left the taproot that were as thick as a cambric needle, and but few smaller ones. These branch roots did not subdivide as rapidly as in most other plants examined.

Washing the clayey subsoil, such as occurred here, from roots as delicate as those of the turnip, would not enable one to gain an accurate idea of their full extent. This probably accounts in part for these findings.

ROOT DEVELOPMENT IN RELATION TO CULTURAL PRACTICE

A study of the very extensive root system with its extremely delicate network of ultimate branchlets explains why turnips thrive best in a deep, rich, moist loam soil and why the soil should be kept in excellent tilth. As with other root crops, such as carrots, beets, and parsnips, a soil that is easily moved by the enlarging fleshy portion of the root is essential; otherwise the roots are very likely to be misshapen and irregular in growth. Heavy clay, for example, crusts of which sometimes form after rains and pinch off the plants, not only increases the difficulty of securing a stand, but root penetration is hindered. Root crops are removed from such soil with considerable difficulty.

The effect of soil environment, as it may be modified by the grower, on the growth of roots of vegetable crops deserves much more study. For example, the effect of phosphates in promoting root growth in length and number of branches has long been recognized in agricultural practice. They stimulate root development in the early stages of plant growth, and under some conditions are used in large quantities in the growth of turnips and rutabagas.

Dressings of phosphates are particularly valuable whenever greater root development is required than the soil conditions normally bring about. Phosphates are needed also for shallow-rooted crops with a short period of growth... Further, they are beneficial wherever drought is likely to set in because they induce the young roots to penetrate rapidly into the moister layers of the soil below the surface.

It would seem reasonable to conclude that the larger the feeding surface at the disposal of the roots the less exhaustive is the crop on the soil, since the food materials would not be gathered from the surface area only. Nitrates, on the other hand, when added to the surface layers of soil, stimulate the production of masses of shallow roots and tend to inhibit root elongation. This would appear to be detrimental to normal crop growth in regions where these layers have very little or no available water during periods of drought, since under these conditions the roots cease to absorb and may die.

The usual spacing of turnips in rows 10 to 18 inches apart and thinning to 6 or even 12 inches in the row result in very considerable root overlapping and underground competition. Shallow tillage conserves the supply of water without root injury and enables the roots to absorb in the surface layer of soil which is usually the richest in nutrients. The large yields often afforded by turnips (600 or more bushels per acre) are produced during only a few weeks of growth. Thus the root system must absorb vigorously to supply water and nutrients to this end.

RUTABAGA

The rutabaga or Swedish turnip (*Brassica napobrassica*) is very closely allied to the common turnip. Both are members of the mustard family and belong to the same genus. One may rather easily distinguish it from the turnip, however, by the short stem or neck at the upper portion of the enlarged stem-root vegetable. It is a smooth-leaved biennial producing a flower stalk 2 to 3 feet tall. Like the turnip it is a cool-season crop but, since it requires more time to mature, it is sown 4 to 6 weeks earlier than the fall crop of turnips.

Seed of the American Purple Top variety was planted June 16 in rows 16 inches apart. Later the plants were thinned to a distance of 5 inches in the row.

EARLY DEVELOPMENT

On July 10 the plants had four large leaves each in addition to about two young ones. Leaves of average size had blades 2.5

inches long by 2 inches in width. The plants had a height of 4.5 inches. Rutabaga is characterized by a taproot, which was at this stage about 3 millimeters in diameter at the soil surface. It tapered to 0.5 millimeter in thickness at a depth of 10 inches. Several roots were traced throughout their rather vertically downward course to their ends at depths of 17 to 21 inches. No branches appeared in the first inch of soil but on a typical plant 19 arose from the second inch, 11 from the third, and 9 from the fourth.

At greater depths branches arose at the rate of four to nine per inch. Figure 8.6 shows the somewhat conical shape of the root system. The longest laterals originated at about the 3-inch level and ran outward but only slightly downward giving the root system a total lateral spread of 12 inches. Most of the laterals, however, did not exceed 2 to 8 inches in length and beyond a depth of 9 inches they were only 1 inch or less long. Branching, in general, occurred at the rate of three to eight branchlets per inch. Usually these were only 0.1 to 1 inch in length. None except the longest had sublaterals.

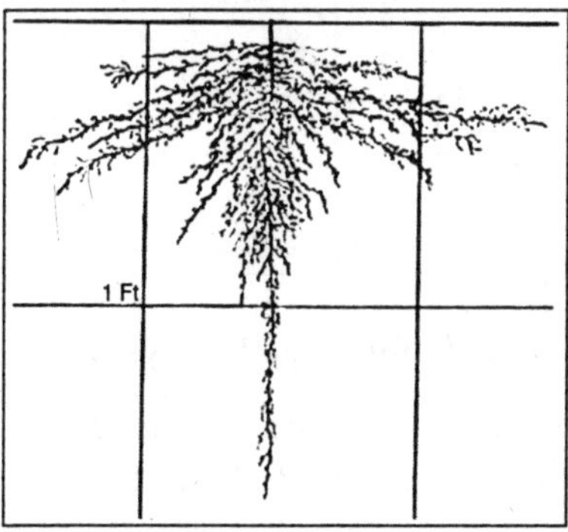

Fig. 8.6 Root habit of American Purple Top rutabaga at the Age of 24 Days.

LATER DEVELOPMENT

Seventeen days later, July 27, a further examination was made. The plants were now 12 inches tall. Each had, on an average, six large leaves with blades about 8 inches long and 5 inches wide. In addition two smaller ones added to the rather large and constantly increasing transpiring area.

The absorbing organs, however, had kept pace in growth with the tops. The taproots had extended downward to depths of about 3 feet. The laterals, in their obliquely outward and downward course, increased the lateral spread to 18 inches. Many of the larger ones had turned downward at various distances from the taproot and, paralleling its course, extended to near and even beyond the 2-foot level (Fig. 8.7). The enlarged portion of the taproots had reached a diameter of 10 to 15 millimeters near the ground line but tapered to only 2 to 3 millimeters in thickness at depths of 3 to 4 inches. Below 10 inches the taproot never exceeded 1 millimeter in thickness and was usually much less.

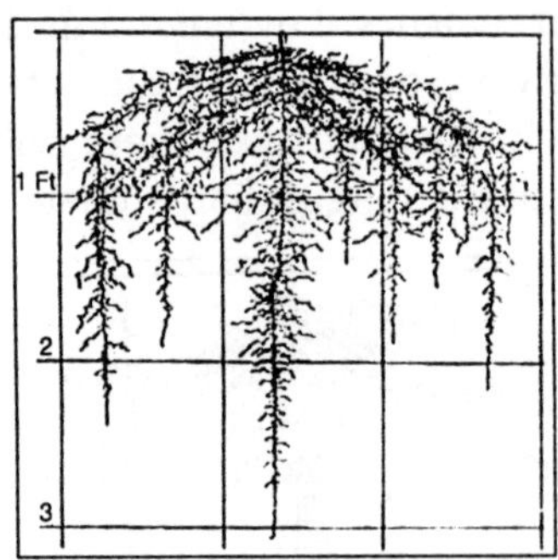

Fig. 8.7 Rutabaga 17 Days Older.

As at the former examination, no branches occurred in the first inch of soil, but frequently 45 to 50 hairlike but densely rebranched laterals were found in two rows, one on each side of the root, in the second inch. These were only 0.5 inch or less in length. At greater depths the number of branches was approximately as before but the branches were larger and longer. Practically all of the long branches, *i.e.,* 9 inches or more in length, arose in the first 8 to 10 inches of soil. Running outward and then downward they filled in the soil volume not occupied by the main root and its branches and thus greatly extended the territory for absorption (Fig. 8.7). Thus the surface 18 inches of soil was well occupied with roots and some extended deeper. Branching was quite profuse, especially near the taproot where 15 to 20 laterals per inch often occurred. Otherwise their former rate of branching (3 to 8 per inch) prevailed. The branches, however, were much longer than before, being 0.2 to 3 inches in length, and all of the larger ones were clothed with laterals of the third order. This added greatly to the absorbing area.

Between depths of 10 and 24 inches the taproot gave rise to six to eight laterals per inch ranging in length from 0.3 to 3 inches. Only the longer ones were rebranched. Beyond 2 feet the laterals were shorter (0.3 to 1 inch) and somewhat less numerous. The last 2 inches of root ends were unbranched.

MATURING PLANTS

By Sept. 5 the plants were approximately 2 feet tall. Each had about 15 large leaves with blades 12 to 15 inches long and 6 to 8 inches wide. The tops had a total spread of about 2 feet.

The lateral spread of roots had not increased beyond that of the preceding examination (about 18 inches). In fact most of the major laterals, which originated in the surface 8 to 10 inches of soil, ran obliquely outward only 4 to 12 inches before turning downward. Aside from the growth of the taproots, which had doubled in length and reached the 6-foot level (maximum, 75 inches), the main extension of the root system was due to the downward penetration of these large laterals. These strong roots varied from 5 to 12 millimeters in diameter and reached depths nearly or quite as great as the taproot.

The fleshy portion of the taproot was about 3 inches thick, tapering to 1 inch in diameter at a depth of 4 inches. It ran in a generally vertically downward direction but deviated from this course through short distances in penetrating the hard soil. Below 3 feet they were often only 1 millimeter in diameter.

From irregular areas on two sides of the fleshy part of the taproot, many fine roots arose. Although very variable in extent, it was not unusual for these areas to cover a portion of the root exterior 2 to 3 inches long and 1 to 2 inches wide. These roots were very much like those on the sugar beet. They were mostly 1 millimeter or less in diameter and from 2 inches to 2 feet in length. With their densely woolly network of branchlets, they formed great masses of rootlets. As many as 80 were counted on a single square inch although where the roots were somewhat larger the number was not so great. A typical root system had 30 major roots, arising in the first foot of soil.

These were intermingled with smaller branches. Many were more than 1 millimeter in diameter and ran outward and downward in such a manner that the soil, 12 to 18 inches on all sides of the plant, was filled with a great network of roots to a depth of 5 feet at least. The degree -of branching can scarcely be overstated. In the first foot of soil the main laterals were clothed with a dense network of branches at the rate of 40 to 45 per inch. All of the branches, with rare exceptions, were thread-like and many were profusely rebranched two or more times.

Below the 10-inch level large branches were rare, never more than 2 or 3 being found. These almost always ran obliquely outward 3 to 5 inches and then pursued a course downward rather parallel to the taproot. Small branches arose at the rate of 10 to 15 per inch although the taproot sometimes ran distances of 2 to 3 inches in hard soil and gave rise to only half a dozen small branches.

In the mellower soil below 4 feet the roots branched much more profusely. Here many of the hairlike laterals were only 8 to 12 inches long and often pursued a course parallel to the taproot. Others spread horizontally or obliquely. With their exceedingly numerous branches they formed cobwebby networks in the moist, mellow soil.

An area of 9 square feet formed one end of the volume of soil occupied, by the roots of a single plant. Of course neighbouring plants also extended their roots into this territory. From just below the soil surface to a depth of at least 20 inches, the soil was a veritable network of rootlets. At greater depths, until the mellow soil at 4 feet was reached, the root branches were confined largely, but not entirely, to the joints in the clay. Here they formed a cobwebby mat of white roots glistening in the dark-coloured soil. Roots were abundant to the 5-foot level and many extended even deeper. Throughout this entire soil volume the root network was of such a pattern that one had to look closely to find the directions of growth of the main roots. Thus because of its great extent and high degree of branching the rutabaga has a very efficient root system.

The rutabaga, closely related to the turnip, has a root habit very similar to it. The rapid growth of the taproot, the wide spread of laterals in the surface foot, and the tendency of those originating deeper to extend more obliquely downward are also characteristic of the turnip. As in the case of the turnip also, the 3 inches of surface soil, until late in the life of the plant, are never well occupied. The vertically downward growth of the ends of the spreading laterals is also the same. Plants 6 weeks old have a lateral spread of 1.5 feet, which is somewhat less than that of the turnip, and rather completely occupy the soil to a depth of 18 inches. The soil volume thus delimited is not extended except in depth. Mature plants have a working level of 5 feet. Branches from the taproot, which penetrate somewhat deeper, though abundant, are not extensive, and the bulk of the absorbing system is formed by the numerous long laterals which originate in the 8 inches of surface soil. In addition, a dense absorbing network of late origin grows near the plant in the soil surface. Throughout the entire extent of the very elaborate root system absorbing masses of rootlets occur in such profusion that their number can scarcely be over-estimated.

9

Horse-Radish and Strawberry

HORSE-RADISH

Horse-radish (*Armoracia rusticana*) is a coarse, hardy perennial with clusters of large leaves somewhat similar to those of the dock. The flower stalks are 2 to 3 feet tall. It is a member of the mustard family. The plants are grown for their roots which are grated or shredded and used as a condiment. A few clumps are found in nearly every home garden where the plants are allowed to grow from year to year and the roots are dug when needed. In commercial gardens it is grown as an annual crop, the plants being propagated from cuttings of the small side roots.

MATURE PLANTS

Vigorously growing horse-radish plants were excavated at Lincoln, Neb., in October. The plants were old, formed large clumps, and had grown undisturbed at least 10 years. The soil was a moderately rich, mellow silt loam grading into a well-drained loessoid subsoil.

The plants were characterized by large fleshy taproots which penetrated deeply but were not widely spreading in their habit of growth. The great clumps of leaves were found to arise from two to several separate stalks which, although quite distinct 1 or 2 inches below the soil surface, were attached to the same root. After several plants had been excavated, a typical specimen was selected for detailed description and drawing.

This plant had a crown consisting of five distinct parts as is shown in Fig. 9.1. The main roots below the crown frequently reached a diameter of 2 inches. The maximum spread of laterals from these roots in the surface foot of soil was 30 inches. Although the average diameter of the soil column occupied by the roots was only about 3 feet, the root system penetrated so deeply that the soil volume ramified by the roots was really very large. Frequently it exceeded 125 cubic feet.

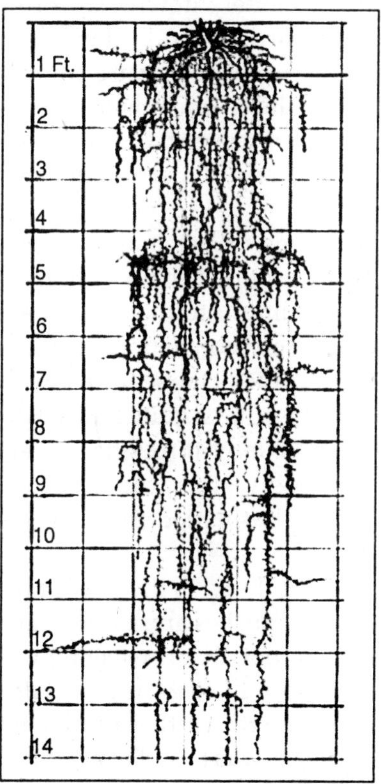

Fig. 9.1 Mature Root System of a 10-year-old Plant of Horse-radish.

Many fine horizontal roots 1 to 12 inches in length arose in the surface soil. Intermingled with these smaller roots were found several roots 1 to 5 millimeters in diameter. They ranged in length from 1 to 3 feet. The larger of these roots turned downward at a distance of 1 foot or less from the base of the plant; the smaller ones usually pursued a more horizontal course.

The main root divided at a depth of 6 to 12 inches into six strong branches, thus reducing the taproot to a diameter of only a few millimeters. In fact it was not as large as some of the main branches, nor did it penetrate so deeply, ending at the 11-foot soil level. The main branches, which were 0.4 to 1 inch in diameter, ran either obliquely downward or outward 3 to 12 inches and then downward. Beyond a depth of 1 to 1.5 feet they pursued a generally vertically downward course, although often curving and turning in the characteristic manner shown in Fig. 9.1. As regards size and appearance, as well as manner of branching, the taproot and the stronger laterals were very similar.

The larger branches of the major roots were fine and numerous and may be grouped either as horizontal or vertical, although rootlets pursuing intermediate directions were not infrequent. Many of the horizontal branches turned downward after pursuing their outward course for a few inches. At levels of 6 to 12 feet where the soil was quite moist, laterals were often found which ran horizontally throughout their entire course. The horizontal laterals were rather uniformly 0.7 to 1.3 millimeters in diameter but varied from a few inches to more than 3 feet in length.

They occurred. irregularly. In the soil above the 6-foot level these larger horizontal branches arose at the rate of one on each inch or two of the main root. Below 6 feet they were sometimes separated by distances of 12 to 24 inches. Although 30 to 40 inches was the usual length of these branches, a maximum lateral spread of 42 inches was found at a depth of 12 feet. Long laterals, however, were not common at this depth. At the 4.5-foot level a layer of hard, lime concretions was found in the soil. Here branches were very abundant. Not infrequently 4 or 5, approximately 1 millimeter in diameter, occurred on only 1 to 2 inches of the main roots. They grew outward a short distance and then turned downward.

More or less vertical major branches of the main roots occurred with less frequency. They were found on all of the main roots, however. These branches usually followed the same course as the main root and not infrequently twined about it, becoming tightly pressed against it as they increased in size. This phenomenon is not shown in the drawing.

The main roots were well covered with fine, absorbing laterals. Their course, as already indicated, was rather tortuous at least to a depth of 9 feet. Here the soil was continuously moist and much more easily penetrated, and the roots responded by pursuing a straighter, downward path. A maximum depth of 15 feet was attained.

The smaller laterals on both the main roots and their major branches were very fine and usually simple, although some were branched. They did not exceed 0.1 to 0.3 millimeter in diameter. Although somewhat irregular in occurrence, they were always abundant, frequently 16 to 24 being found on each inch of main root. In the first 3 feet of soil this number was often exceeded. Here it was not unusual for the branches to occur in clusters of 2 or 3. The number of laterals on the small branches and on the main roots at depths greater than about 6 feet was 16 or less per inch. These absorbing roots were frequently 3 to 4 inches in length in the first 8 feet of soil. But the root length rather gradually decreased from the surface-soil layers to within a few inches of the root tips where rootlets had not yet formed. At a depth of 12 feet they averaged only about 1 inch in length. Branches were also shorter on the smaller laterals. The roots were white in colour and could be easily identified at any depth by their characteristic taste.

Horse-radish is a perennial with a very thick, fleshy taproot system which penetrates to depths of 10 to 14 feet but does not spread widely. The taproot gives rise to many fleshy laterals in the surface foot of soil, where also frequently the main root divides into several rather equally prominent branches. These penetrate more or less vertically downward, or, more usually, first run obliquely outward 4 to 12 inches and then pursue their very tortuous downward course. A diameter of 4 millimeters is often retained even at a depth of 6 feet. Supplemented in the surface soil by numerous fine horizontal roots and profusely branched throughout, the main roots ramify a very large soil volume. The larger of the fine laterals are usually either horizontal or vertical in direction of growth and are from a few inches to over 3 feet in length. Their distribution is somewhat irregular. Smaller branches are numerous. Thus the absorbing system as a whole is both extensive and profuse.

ROOT HABIT IN RELATION TO CULTURAL PRACTICE

The great extent and rapid development of the horse-radish roots explain why the plants grow best in a very deep, moist, fertile loam soil; why deep plowing is beneficial; and why a good soil structure is an important environmental factor for growth. Soils for vegetables should always be deep. A depth of 8 to 12 inches is desirable for most soils. But if they are shallow, they should be deepened gradually and not all at one plowing or in a single year. Too much subsoil brought to the surface at once is usually not beneficial. In planting the roots, they are placed with the upper end of the cutting 2 to 5 inches below the soil surface. The soil should be packed firmly about them to insure good contact and resultant prompt growth.

Weeds should be kept out by thorough cultivation so that sufficient water and nutrients will be available late in the season when the plants make their best growth. On hard, dry, or shallow soil the roots are very likely to be crooked, unshapely, and scarcely fit for use. The growth of long, straight roots of more uniform size is promoted by preparing and maintaining a deep, mellow soil. To aid in securing shapely roots... some growers remove the side roots early in the season. This is done by removing the soil and stripping the side roots from the upper part of the main root. The soil is then replaced. This treatment results in the production of large, compact roots, but unless the work is carefully done, serious injury may follow. The earlier in the season the trimming is done the less check there is to growth.

Since plants several years old have a lateral spread of roots scarcely exceeding 18 inches, it would seem that the usual spacing for plants grown a single year is sufficiently great to allow a good root development. The usual distance is 10 to 18 inches in rows 3 to 4 feet apart. If allowed to grow more than 1 year, the roots are apt to become hollow and undesirable for market.

RADISH

The radish (*Raphanus sativus*) is one of the commonest of garden vegetables, being especially in favour with the home gardener because of the ease and rapidity with which an early spring crop may be obtained. If planted early, it forms seed and completes its life cycle in a single season. Varieties planted later set seed during the second season of growth. The branched flower stalks are 1 to 3 feet in height. The plants are grown for their pungently flavored roots, including the upper portion of this swollen, edible part which is morphologically a stem. The "roots" vary much in size and shape from those that are short and globular, through conical and oblong, to the spindle-shaped type.

EARLY SCARLET TURNIP WHITE-TIPPED RADISH

Seed of the Early Scarlet Turnip White-tipped radish was planted Apr. 10. The rows were 14 inches distant and the plants thinned to 2 inches and later to 4 inches apart in the row.

Early Development

Have been observed to form six to eight laterals on the upper inch of the taproot only 4 days after planting and at a time when the cotyledons were just unfolding. Nine days later, when the plumule was appearing, secondary laterals had been formed. This shows that a plant needs an efficient absorbing system before much leaf area is exposed to the dry air.

On May 18 the plants each had about five leaves and other leaves were just appearing. The average leaf surface was 22 square inches. The total spread of tops was about 6 inches.

The fleshy portion of the strong taproot was 1 to 1.5 inches in diameter. The taproot pursued a nearly vertically downward course, curving laterally through short distances. An average depth of 19 inches and a maximum penetration of 22 inches were reached. The diameter of 2.5 millimeters, found just beneath the enlarged portion of the taproot, decreased at a depth of 6 inches to only 1 millimeter, a thickness that was maintained throughout the course of the root. Just below the enlarged part of the taproot, or at most only a few millimeters below, numerous branches arose (Fig. 9.2). These extended outward in a generally horizontal direction but often curved upward or downward. The longest reached a distance of 12 inches from the base of the plant, but most of the roots were only 2.5 to 3 inches in length.

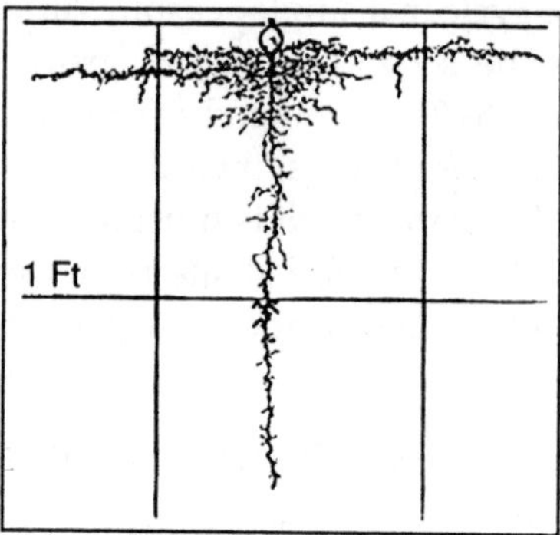

Fig. 9.2 Root System of the Early Scarlet Turnip White-tipped Radish about 5 weeks Old.

Laterals were thickest on the first 3 inches of the taproot, where they occurred often in groups of twos or threes, in two more or less distinct rows on opposite sides of the taproot. From 40 to 45 roots often sprang from the first 3 inches of the slender portion of the taproot. Below this they were fewer, shorter, and more poorly branched or entirely destitute of laterals. They, spread horizontally on all sides of the plants. Branching was very profuse. Laterals of the first order occurred at the rate of 5 to 20 per inch and varied from 1 millimeter to 2 inches in length.

Although branches of the second order were found but rarely, the thread-like roots of these young plants thoroughly ramified the surface layer of soil.

Effect of Soil Structure on Root Development

Radishes were grown in rectangular containers with a capacity of 2 cubic feet and a cross-sectional area of 1 square foot. A rich, sandy loam soil of optimum water content was screened and thus well aerated. One container was filled with soil with very slight compacting. It held 173 pounds. Into the second container, 232 pounds of the soil were compacted. Surface evaporation was reduced by means of a thin sand mulch. When the plants were 3 weeks old and five or six leaves had developed, the side of each container was cut away and the root system examined. The taproots had reached a maximum depth of 22 inches in the loose but only 5 to 8 inches in the compact soil. The laterals in the first 2 inches of dense soil were more numerous, of somewhat greater diameter, and slightly longer (maximum 6.5 inches). They were also much more branched. But in the loose soil branching occurred throughout to near the root tip. That roots of many species penetrate more deeply in loose than in compact soils and that under the latter condition branching is increased has been shown by numerous investigations.

Two-months-old Plants

The plants now had stems nearly 2 feet tall. with 9 to 11 branches. They were flowering abundantly. A few of the oldest leaves had turned yellow and dried. The cauline leaves were relatively smaller than the basal ones. Plants of average size had leaf surfaces slightly less than 3 square feet in extent.

At this time the fleshy portion of the root was about 2.5 inches thick. On about half the plants a second storage reservoir was being formed below the first, or at least the root was considerably thickened. The strong taproot, about 4 millimeters thick throughout the first 8 inches, which was the most branched part of its course, ran quite vertically downward to a depth of 3 feet. The root ends were not thick as at the preceding examination but greatly attenuated and quite free from branches for distances of 5 to 7 inches. In fact below 10 inches branching was relatively sparse although greatly promoted where the roots entered earthworm burrows or soil areas enriched by the presence of former roots. Here they sometimes coiled about and branched much more profusely.

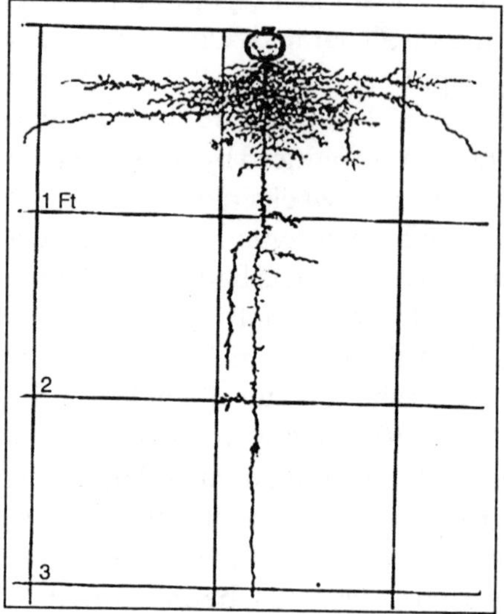

Fig. 9.3 Characteristic Root Habit of a 2-months-old Radish. Most of the Fibrous Roots are Distributed in the Surface Soil.

A remarkable degree of branching occurred between the depths of 2 and 8 inches. In one representative plant 49 laterals arose from the first inch of taproot below the enlarged part, 34 from the second inch, 20 from the third, and 11 to 15 from each of the next 3 inches, respectively. Below this, in the first foot, the taproot branched at the rate of 5 to 7 laterals per inch. Many of the branches arose in groups of twos or threes from two rows on opposite sides of the root. They usually extended horizontally in all directions from the taproot. Many were only 2 to 4 inches long; some reached distances of 13 to 17 inches from the base of the plant. That they, like the taproot, were growing rapidly was clearly indicated by the long, unbranched root ends. The laterals were well furnished with branches which occurred at the rate of 4 to 10 per inch. Although many of these were very short (0.2 inch or less), others reached a length of 1 to 3 inches. Most of the older ones were rebranched with short laterals and branches of the third order were frequent. The thread-like character of the branches and the great network of delicate branchlets are very characteristic of the radish and make it quite difficult to excavate.

Mature Plants

A final examination was made July 13 when the plants were 26 inches tall and in the flowering stage. The vigorous plants had 18 to 22 leaves, usually with a flower stalk in the axil of each. The large transpiring surface (nearly 5 square feet) may be visualized when it is recalled that the larger leaves were 8 to 12 inches long and 4 to 6 inches in width. Just below the soil surface the enlarged globular portions of the fleshy taproots were 2.5 to 4 inches in diameter. Although usually cracked and fissured, they were still quite firm and well stored with food. Below this portion secondary, irregular enlargements occurred on practically all of the many plants examined. These varied from 0.5 inch to over 2 inches in diameter, the taproot frequently being swollen for a distance of 6 to 8 inches.

Even a casual inspection of Fig.9.4 shows the marked growth the roots had made since the examination a month earlier. The taproots frequently reached depths of 6 to 7 feet, and a maximum penetration of 7.2 feet was found. The widely spreading, horizontal lateral branches had extended the absorbing area to over 3 feet on all sides of the plant. A maximum lateral spread of 4.1 feet was ascertained. Not only was the surface soil from 3 to 12 inches thoroughly ramified over a wide area but also the deeper soil, even beyond 4 feet, was penetrated by strong, well-branched lateral roots.

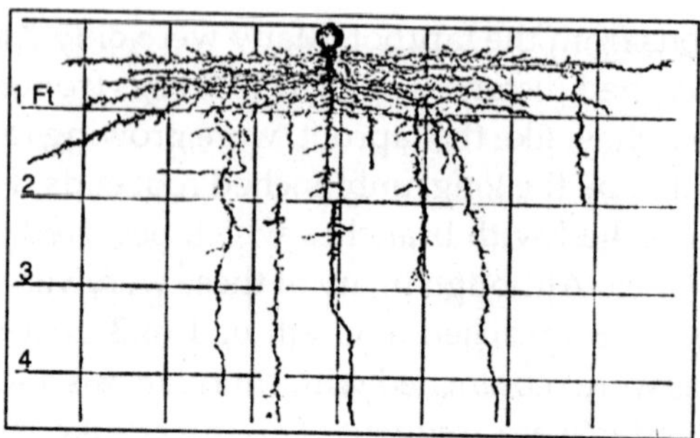

Fig. 9.4 Root System of Radish that had Formed a Flower Stalk, Excavated July 13. The Insert (A) is a Continuation of the Taproot which Reached a Depth of more than 7 Feet.

The taproots pursued a generally vertically downward path but with characteristic kinks and curves as indicated in the drawing. Throughout their course in the surface 4 to 12 inches of soil they were so covered with branches, both large and small, that they formed a cobwebby network. Below 12 inches, the taproots, now about 2 millimeters in diameter, were very much kinked and curved and sometimes turned abruptly. Large laterals were not abundant, frequently only 5 to 8 occurring below the first foot of soil.

These were rather poorly furnished with short branches and seldom exceeded 8 inches in length. Their direction of growth was mostly horizontal. No branches over 1 to 2 inches long occurred at greater depths, but numerous, short branches were found throughout the course of the taproot.

The larger lateral roots, originating in the surface 4 to 12 inches of soil, usually occurred at the rate of six or more per plant. They were 1.5 to 2 millimeters in diameter throughout the first 8 to 12 inches of their course which, as among the very numerous smaller laterals, was characterized by many kinks and turns. Frequently, the main laterals gave rise to branches quite as large as themselves. All branched profusely, frequently to the third and fourth order, especially the older parts. Portions of the laterals were found that had only three to four branches per inch or, indeed, were destitute of them; they were typically clothed with rootlets 0.4 to 1 inch long at the rate of 8 to 12 per inch. As a whole the root system of the radish is rather clearly differentiated into a deeply penetrating portion consisting of the taproot and a few of its widely spreading and deeply penetrating branches and a shallower part which spreads widely but is almost confined to the surface 4 to 12 inches of soil. Here dense masses of delicate, cobwebby roots were profuse. This soil layer was so thoroughly exhausted of its water that during drought many of the cobwebby rootlets of the intricately branched network of roots shriveled and died. The absence of roots in the surface few inches of soil is characteristic.

Summary

The Early Scarlet Turnip White-tipped radish is characterized by a rapidly growing and deeply penetrating taproot. At the time of. the removal of the fleshy portion of the root for table use, the root system is not extensive when compared with that of most vegetables. Although it may penetrate to a depth of 2 to 3 feet and have a lateral spread of 12 to 16 inches, most of the absorbing area lies in the surface 2 to 8 inches of soil.

Even on fully matured plants only the portion of the taproot lying in the 2 to 12 inches of surface soil gives rise to large laterals. Most of these extend horizontally to distances of 3 feet or more on all sides of the plant. They branch profusely and rather thoroughly fill the surface soil with a network of absorbing rootlets. The deeper soil is ramified by the taproot and by a few of the major roots or their branches which originate in the surface foot. The taproot penetrates to depths of 6 to 7 feet. The large branches reach depths of 3 to 4 feet and are well clothed with short laterals.

THE EARLY LONG SCARLET RADISH

Seed of the Early Long Scarlet radish was planted at Norman, Okla., Apr. 18. Conditions for growth were very favourable both as regards temperature and soil moisture and the plants grew vigorously. When 20 days old and at a time when the tops consisted of five to seven leaves, the first root examination was made.

EARLY DEVELOPMENT

The taproots, some of which were already 0.5 inch in diameter, had grown at the rate of 1.2 inches per day. They reached a maximum depth of 2 feet. Branches were found from just beneath the soil surface to near the root ends. They were very abundant, 37 to 53 occurring on the first foot of the taproot alone. In length, they ranged from a few millimeters to 15 inches. Their direction of growth was mostly horizontal although below 8 inches a few ran obliquely outward and downward. All of the longer ones were branched. Branches occurred at the rate of about 4 per inch and many of them were 1 to 2 inches long. These also were rebranched.

The entire root system was covered with root hairs and was apparently functioning vigorously. Most of the absorbing surface occurred in the first foot of soil, some of the laterals extending almost to the soil surface. But the second foot was also ramified by the taproot, by a few obliquely descending major laterals, and by the rather numerous branches from these.. The second foot of soil was moist and fairly mellow and was soon to become the seat of vigorous root activity. At this time the root system had, in general, the shape of an inverted cone 2 feet long and about 28 inches wide.

LATER DEVELOPMENT

Twelve days later, May 20, the fleshy portion of the taproot had increased to 1 inch in diameter. The roots were enlarged to a depth of about 8 inches. Some of the major branches were also fleshy near their origin. Marked changes had occurred during the 12-day interval. The depth of penetration had increased to 3 feet and the lateral spread to a maximum of 2 feet. Although the general shape of the root system had not changed, many of the secondary laterals had greatly elongated; the absorbing area was much extended; the second foot was well ramified to a distance of 8 inches on all sides of the taproot, and the third foot was almost as fully occupied as was the second 12 days earlier. Branching throughout was even more abundant. The root system was growing vigorously.

MATURE PLANTS

The plants continued their rapid growth throughout June and by July 6 they averaged 30 inches in height. The tops spread so widely that they nearly covered the soil, although they had been thinned earlier until the mature plants were 18 to 24 inches apart in rows 3.5 feet distant. Blossoming had ceased and the seed pods were maturing.

The fleshy roots were 3 to 5 inches in diameter and 8 to 14 inches long. The taproots were several millimeters in diameter, however, even to a depth of 3 feet. They pursued a generally downward, although somewhat tortuous, course through the rather compact subsoil and reached depths of 4.5 to 5 feet. Several of the major laterals, originating from the fleshy portion of the root or near its base at depths of 8 to 10 inches, had a lateral spread of 40 inches. Many of these ran far outward and then turning downward penetrated to depths of 2 to 3 feet. Others pursued a horizontal course throughout but gave rise to long, vertically descending branches some of which reached the 4-foot level. Lateral branches on all of the roots were very much longer than formerly, penetrating the soil in all directions for distances of 6 to over 30 inches.

In the process of growth expansion the plants had forced themselves 1 or 2 inches out of the soil. Similar root heaving has been observed on other fleshy rooted plants. 83 This was clearly shown not only by the exposed crown and upper portion of the fleshy root but also by the abrupt inward curving of the main roots originating from the fleshy part of the taproot. Earlier examinations showed this same type of root running horizontally. The phenomenon was very pronounced since numerous curved branches 2 to 10 millimeters thick all turned obliquely downward at their origin. Some of the shallower horizontal roots were actually stripped of their laterals and pulled above the soil surface.

Owing to the very dry soil and perhaps in part to high temperatures, many of the roots in the shallower soil had ceased absorbing; others had died. Very few active roots occurred within a radius of 2 feet from the base of the plant and to a depth of over 12 inches. Thus absorption was confined to the younger parts of the horizontal branches and to those lying deeper in the soil. The root system was very extensive, however, both in lateral spread and depth thus ramifying a very large soil volume. Throughout this soil mass, except as already indicated, branching was very profuse; great mats of glistening white rootlets filled the soil. It is remarkable that a plant in such a short period of growth can produce such an extensive and intricate absorbing system.

SUMMARY

The root system of the Early Long Scarlet radish is of an entirely different type from that of the Early Scarlet Turnip White-tipped variety. In its early development the vertically penetrating taproot was branched throughout, mostly with horizontal laterals, from the soil surface to near its tip, the root system having the general shape of an inverted cone. When the roots have reached the size used for eating, they penetrate to depths of 3 feet and the much-branched laterals spread radially to distances of 2 feet, the conical nature of the root system being in general retained. Mature plants are characterized by very fleshy taproots, 4.5 to 5 feet long and many major laterals, all of which arise in the surface foot of soil. These spread widely (maximum, 40 inches) and often turn downward near their ends. They also give rise to major laterals which penetrate deeply. Thus, although the surface soil is well occupied, the subsoil to depths of 3 to 4 feet is also more or less filled with the much-branched lateral rootlets. Further investigation may show that these differences in root habit are due in part to varietal characters and in part to soil environment.

OTHER INVESTIGATIONS ON RADISH

The root systems of the Gray Summer Turnip variety and

the London Particular Long Scarlet were washed from the soil at Geneva, N. Y. The roots of both penetrated the soil a distance of 2 feet and the branches extended on either side more than 21 inches, mingling with those from adjoining rows. The taproot did not begin to branch much until some distance below the edible part. The branches at first were few in number, usually but two or three from the taproot. These extended nearly horizontally, and ramified towards their extremities into many fibrous roots. The greater part of the feeding roots lay in the upper 8 inches of the soil. Though the edible roots of these two varieties are quite different in f orm, their rooting habits show no difference. Certain German investigators have also observed that the fleshy portion of the main root is practically without branches; that the taproots reach depths of 12 to 20 inches; and that the ultimate branching is very diffuse, branches of the third order being abundant. Steaming the soil has been found to increase the total number, size, and yield of radishes, a fact which is undoubtedly connected with a greater development of the root system.

ROOT DEVELOPMENT IN RELATION TO CULTURAL PRACTICE

Although the radish, like its wild ancestors, will grow in nearly all types of soil, a light, friable) fertile soil is best. Since the crop develops very rapidly and the root system, except in plants grown for seed, is not extensive, the most favourable conditions for root activity should be attained by thorough seed-bed preparation and shallow cultivation. Stirring the 2 inches of surface soil will not harm the roots, since practically all of them lie deeper. They do nearly all of their absorbing in the surface 6 to 8 inches. Hence, water should be conserved in this soil layer, manure and fertilizers supplied to it abundantly, and competing weeds kept out so that growth will be rapid and continuous.

It is well known that radishes will thrive under close planting or as an intercrop between rows of later-maturing plants. A knowledge of the shallow, non-extensive, but well-branched root system helps one to understand why this is the case. The ramification of the soil by the roots of older plants and probably by longer-season varieties is quite a different matter. Undoubtedly close planting, which results in shading aboveground as well as root competition in the soil, affects the size and development of the fibrous portion of the root system just as it does the edible part. The effect of shading on the growth of roots has been clearly demonstrated, for example, in the case of certain tree seedlings.

In White-pine Seedlings Grown in Vermont

Darkness induces the growth of tall seedlings with poorly developed roots; a diffuse light, the growth of shorter plants and longer roots; and the full light produces short stocky plants with long branching roots.

For example, seedlings of similar age grown in the nursery under full shade had unbranched taproots 3.5 centimeters long, those grown in half shade were 4.5 centimeters long and had the beginnings of laterals, And seedlings grown in full light had taproots 5.2 centimeters long and a lateral development of roots nearly seven times as great as those in half shade. Among 50 seedlings excavated about 3 weeks later, the length of the root systems were 4.5, 8.2, and 13.8 centimeters and the total number of lateral branches on the lot 5, 143, and 468, respectively.

Plants that are crowded cannot develop efficient root systems and their activities are confined to the shallower soil. Unless this is kept very rich and well moist, drought and reduced yields are inevitable. Since plants with like root systems are making similar demands at the same depth at the same time, they are least fitted to be crowded together. Intercropping and growing mixed cultures more nearly approaches nature's method of producing a dense growth of vegetation. For scientific progress in this direction a thorough understanding of root relations is indispensable.

STRAWBERRY

The strawberry (*Fragaria chiloensis*) is an herbaceous

perennial. The very short, thick stems occur close to the surface of the soil. Although not always included among vegetable crops, its wide distribution, the temporary nature of the crop, and its common occurrence in gardens warrant its inclusion here. It is the most valuable of the small fruits grown in the United States. It is indeed a cosmopolitan plant, growing and thriving under a wide range of conditions. The best crops are usually grown in rich, rather moist soil during a cool season.

MATURE PLANTS

A strawberry of the Dunlap variety, one of the most important commercially, was studied on an area of silt loam soil about a mile distant from the experimental field in Lincoln. The plants were excavated June 24 when blossoming and fruiting were nearly completed. They had grown 3 or 4 years in this field.

The underground parts are characterized by a dense network of fibrous roots which arise from the much-thickened, short, scale-covered stern. Plants of average size possessed about 36 roots, each approximately 1 millimeter in diameter, and about 9 smaller ones. These fibrous roots radiated outward or downward in all directions from the horizontal to the vertical. Many spread outward somewhat obliquely and turned downward.

A maximum lateral spread of 12 inches and a maximum depth of 37 inches were ascertained. The root system was characterized by a dense network of fibrous roots in the 10 to 12 inches of soil next to the surface. In the surface foot, and especially in the upper 6 inches, the roots were rebranched at the rate of 3 to 10 fine rootlets per inch. The branches were usually quite kinky And varied from 0.3 to 6 or even up to 10 inches in length.

The longer ones usually- occurred on the older roots. The primary branches were rebranched at the rate of 3 to 8 per inch with laterals 0.1 to 1 inch long. The longer secondary laterals were again rebranched. Where the main root had been cut or otherwise injured, 3 to 5 long branches frequently arose from the root tip. Such branches usually extended only slightly horizontally but nearly always ran rather directly downward.

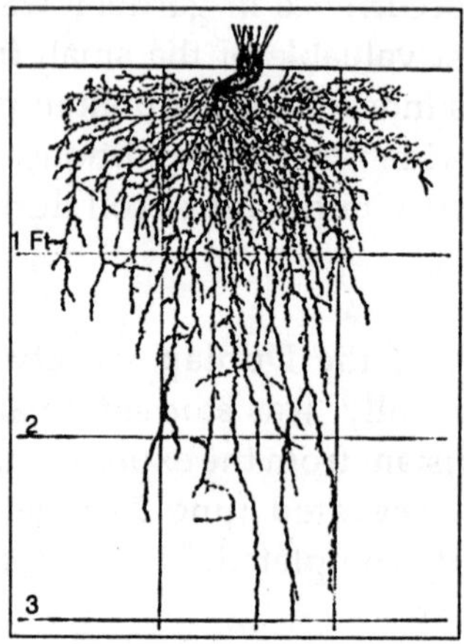

Fig. 9.5 Underground Parts of a 3-year-old Dunlap Strawberry Excavated in June after Blossoming and Fruiting.

Below the surface foot of soil the rate of branching was somewhat less, usually 3 to 6 rootlets per inch, and the branches were shorter, although rootlets 3 to 4 inches long were frequently found. The first 2 inches of roots near the stem were branched but little. Occasionally a root 1.5 millimeters in diameter, originating from the rhizome, was more profusely branched than the rest, the dense branching being largely confined to the surface foot of soil.

The young roots are white in colour. As they become older, their colour changes to a Fight brown and finally to very dark brown or nearly black. Plants starting from runners have many long, young, light-coloured roots.

SUMMARY

The fibrous root system of the strawberry arises from the short, thick stems near the soil surface. Just beneath the surface horizontal roots extend about 1 foot on all sides of the base of the plant. This delimits the lateral spread. The surface foot of soil is fully ramified by obliquely descending roots as well as many more or less vertically descending ones. The latter, especially, also ramifies the second and some of them the third foot of soil. Branches are mostly short but abundant, usually being more profuse in the surface 12 inches. The root system is relatively shallow and not extensive.

OTHER INVESTIGATIONS ON STRAWBERRIES

A plant of Triomphe de Grand variety of strawberry was washed from the soil at Geneva, N. Y., in midsummer. The horizontal roots were found to be few and short, the longest being traceable only 6 inches. The greater part of the roots extended nearly perpendicularly downward, and nearly, all the fibrous roots were found directly beneath the plant. The roots reached a depth of 22 inches. New roots were found growing from the rhizome about 1 inch above the old ones. The longest of these had attained a length of 6 inches. They were white and tipped at their extremities with a thickened point.

Strawberries of the Warfield variety were studied at Madison, Wis. The plants were 3 years old. They were washed from the soil late in May when the fruit was maturing. The roots were contained within a very small soil volume. The deepest roots extended a little less than 2 feet and the horizontal ones reached scarcely beyond the area covered by the leaves. Most of the roots grew downward and all but the merest fraction of them were contained within the first foot of soil. The soil was a light, clay loam which was underlaid at a depth of 1 foot with a subsoil of sandy clay. The soil had been cultivated about 3 inches deep.

It was found at the same station that certain herbaceous plants, including the strawberry, start growth extremely early in spring. As early as Mar. 22, when the ground was free from frost in places receiving the most insolation, the roots of strawberry had made considerable growth. Root growth in spring is most active near the surface of the soil which is first warmed by the sun's rays. "Thus the parts of the soil from which roots are in a measure excluded during the dry weather of summer may serve as a feeding ground for them in early spring." That there is considerable difference in the root habits of varieties is indicated by the statement that some varieties which are usually a failure in dry climates because of their deficient root system are enabled by irrigation to flourish to such a degree as to be among the most profitable. The root system of the wild strawberry (*Fragaria virginiana*) has been examined where the plants were growing in a spruce forest. The lateral spread and depth of penetration were very similar to that described and illustrated for the cultivated variety. In drier habitats it would undoubtedly be more extensive.

ROOT HABITS IN RELATION TO CULTURAL PRACTICE

The fact that native strawberries, from which the cultivated varieties have arisen, grow in nearly all types of soil from seashore to mountain top, indicates a wide range of soil and climatic conditions for cultivated plants. The rather limited root range of the plant easily explains its well-known susceptibility to drought as well as its ready response to fertilizers.

It emphasizes the importance of providing irrigation in regions of moderate and fluctuating rainfall if large and regular crops are to be obtained. The need of a rather constant water supply about the roots is indicated by the distribution of native strawberries in dry grassland areas. Here they are found in the better-watered soils in ravines and along creeks and river bottoms, and often fringe thickets and woodlands. Perhaps an ideal soil would be a sandy or gravelly loam underlaid with a pervious clay, *i.e.*, one retentive of moisture yet easily tilled. By proper methods of tillage, mulching, and irrigation much can be done towards furnishing the roots with a constant moisture supply. Soil for strawberries should be thoroughly prepared. Deep plowing and thorough pulverizing a considerable time before transplanting encourages the roots to penetrate more deeply, thus lessening the danger from drought and injury from cold. Soil thus prepared will catch and retain more moisture than a poorly and recently plowed soil. In transplanting, only large, sturdy, young plants with a vigorous root system should be selected. These plants have thick, turgid, light-coloured roots quite in contrast to the dark, less vigorous appearing roots of older ones. A plant with a healthy root system and a comparatively small top is much to be preferred to one in which these conditions are reversed.

The transplants make a better growth if both tops and roots are pruned before setting. Very little pruning of the tops will be required if the plants are secured early. The roots are frequently cut back to a length of 3 to 5 inches. The removal of a portion of the root system permits better spreading of the roots and facilitates transplanting. Great care should be exercised so that the roots do not dry out in the process of transplanting.

It is also important to spread the roots in the soil and to press the moist soil firmly against them, thus establishing good contact. Care should also be taken to set the plants at the proper depth. When the roots are too deep and the stern is buried under the soil the crown or terminal bud will be covered with soil and the plant may not grow. Exposing too much of the stem, thus placing the roots too near the surface where they will become dry, is also harmful (Fig. 9.6). The fact that new branches from the perennial stem appear above the older ones explains the tendency for the short stern to become more and more above the soil as the plant becomes older.

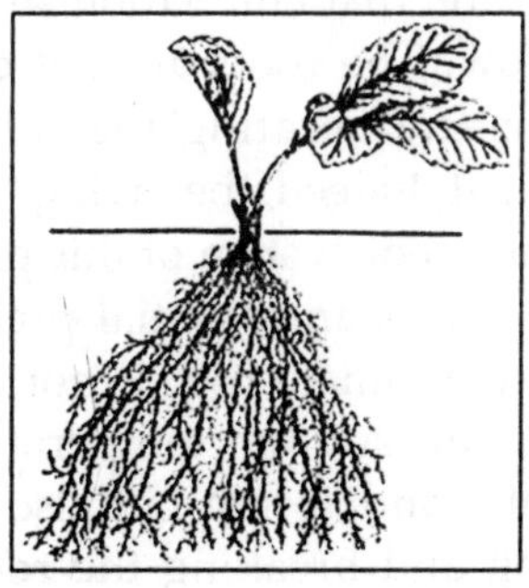

Fig. 9.6 A Strawberry Plant Correctly Pruned and Set at the Proper Depth.

Because of the method of propagation by rooting aboveground stems (stolons or runners), commercial growers generally prefer the matted-row system of planting since it is the simplest and easiest to maintain. Moreover, it almost invariably gives the highest yields. In one experiment 43 varieties were employed. The transplants are often set 10 to 30 inches apart in rows 3 to 3.5 feet distant.

The runners are allowed to form plants 6 to 12 inches on either side of the row thus leaving sufficient space between the rows for cultivation. Sometimes the offshoots are spaced 6 toll inches apart. Under such conditions the roots thoroughly ramify all of the soil and extend a considerable distance between the matted rows. In fact all of the space between the rows may he well occupied by roots. They are disturbed by cultivation unless it is very shallow. Plants in hills or hedge rows often develop more extensively, according to the root and shoot competition imposed upon them, and the fruits are usually larger.

Proper cultivation is said to be the most important factor in strawberry production. The root system needs thorough aeration and a constant supply of moisture. Both of these conditions. are attained by preventing the formation of a soil crust due to driving rains or irrigation. Keeping the soil mellow also encourages the rooting of the runners. Strawberries will not stand competition of weeds; the tops are easily overshadowed and the shallow roots are relatively not extensive. It is important, therefore, that cultivation should be thorough and timely but shallow. The roots are not exhaustive of soil nutrients. Loosening and aerating the soil seems quite as important as manuring it. Indeed, the saying "tillage is manure" applies very well to the root system of this plant.

Mulching in the fall is an essential practice with such a superficially rooted, perennial crop. It not only protects the aboveground parts from sudden temperature changes and drying winds in winter and spring but also tends to prevent the soil from heaving and breaking the roots, Moreover, it conserves moisture and inhibits the growth of weeds. By leaving the mulch intact in the spring the soil is kept cool and damp during the season when the fruit is being produced. Because of the lower soil temperature, absorption and growth are slower and blossoming is delayed. The root relations also explain why a mulch is usually more beneficial in regions of light and precarious snowfall than in those in which a blanket of snow lies on the ground all winter.

10

Effect of Tropical Vegetation on Seeds

ROLE OF GARDEN PEA SEEDS

The garden pea (*Pisum sativum*) is a hardy, cool-season annual. It is quite variable in size, ranging from 1 to 6 feet in height, since some varieties are dwarf, others half dwarf, and still others tall. It is cultivated in nearly all home gardens, in market and truck gardens, and is also grown on a very large scale for canning. Since it does not thrive during midsummer it is grown as a partial-season crop, *i.e.*, in spring and early summer in the North and as a winter and spring crop in the South. Like other members of its family (*Leguminosae*), through the agency of bacteria in its root nodules, the pea is able to utilize nitrogen compounds which have been formed by using the free nitrogen of the soil air.

Garden peas of a late or main-crop variety, known as Telephone, were planted Apr. 10. They were sown in rows 30 inches distant and the plants were later thinned to 8 inches apart in the row.

EARLY DEVELOPMENT

The plants were excavated for the first time on May 23. They were 11 inches high and the stems 5 millimeters thick. Each plant had two or three branches and about a dozen large compound leaves some of which were not fully developed. The total leaf surface was 1.5 square feet.

The plants had strong taproots about 4 millimeters in diameter at their origin. These tapered to 1 millimeter in thickness below the 9-inch level. Depths of penetration of 20 to 26 inches were found. The course of the roots was often somewhat tortuous. An examination of Fig. 10.1 shows that the bulk of the absorbing area was furnished by the numerous, strong, lateral branches nearly all of which arose from the taproot in the surface soil. Just below the seed, which was 1 inch deep, these occurred in great abundance. One rather large plant gave rise to 21 branches about 1 millimeter in diameter from the first inch; 14 nearly 1 millimeter thick from the second inch; 11 averaging ½ millimeter in diameter from the third inch; and 16 more, about 1/3 millimeter thick, in the next 6 inches. The specimen drawn had only about two-thirds as many roots.

The general course of these roots was outward and only slightly downward. Thus nearly all ended in the first 6 to 9 inches of soil. A few of the longest extended laterally 15 to 18 inches and then turned downward. Depth of penetration, however, was not marked and only a few reached the 12- to 18-inch soil level. Hence, the plant had a rather superficial root system at this stage of growth. The absence of roots in the surface 1 to 2 inches of soil is of interest since it bears a direct relation to root injury by cultivation.

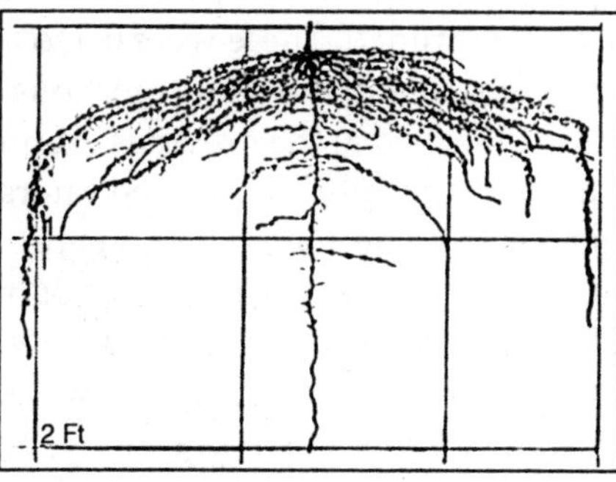

Fig. 10.1 Root System o f the Telephone Garden Pea 6 Weeks Old.

As regards branching of the taproot, it may be noted that some of the laterals, even in the surface 6 inches, were only 3 to 9 inches long. Below this depth branching was much poorer as is clearly illustrated in the drawing. The length of the laterals decreased with depth of origin and the last 4 to 6 inches of the taproot were unbranched. This, of course, was also true of the major laterals. These were furnished with rootlets at the rate of two to eight per inch, distribution being somewhat variable. These sublaterals usually varied from 0.2 to 1 inch in length, although they were occasionally longer. They were rarely branched.

LATER DEVELOPMENT

A second examination was made June 17 when the peas were blossoming. There were about four branches per plant and these varied from 1 to 2.5 feet in length with an average length of 18 inches. Plants of average size possessed 45 leaves and a total leaf surface of slightly more than 5 square feet.

The pronounced taproots had increased to 7 millimeters in diameter near the soil surface. They tapered to 1.5 millimeters at a depth of 1 foot but were nearly 1 millimeter thick throughout their sinuous course. Maximum depths of 36 to 38 inches were ascertained. Branching had increased to a marked degree; a few roots 0.5 to 1 inch long now occurred in the first inch of soil. Frequently as many as 100 laterals were found to originate from a taproot in the first 12 inches of soil. A maximum of 18 per inch was determined. These were 0.5 to 2.5 millimeters in diameter. In the second and third foot-portions of the taproot which were poorly branched at the previous examination-a total of 110 to 130 rather uniformly distributed laterals frequently arose. The branches become longer, in general, on the older portions of the taproots. Near the root ends no branching occurred. The lateral spread had increased only a few inches beyond that of the previous examination (now about 22 inches) but many of the widely spreading roots had turned downward and extended well into the second foot of soil. Moreover, the branching had progressed with the growth of the main laterals.

Branches occurred at a somewhat variable rate, about seven per inch of lateral. This was also the usual rate of branching for the main laterals. Usually the branches were only 0.2 to 0.5 inch long although a few of them attained a length of 2 inches. Thus new soil areas were ramified for water and nutrients.

A pronounced feature was the greater number of branches on the portions of the laterals at some distance from their origin. In fact the first few inches were often poorly branched. Branchlets of the third order were more pronounced than at the previous examination although on the newer growth only the longest secondary laterals were rebranched. No long laterals arose below the first foot of soil but the very abundant short ones (1 to 3 inches) added considerably to the absorbing area.

In summarizing, four differences were apparent from the previous root development: The taproots had increased in depth from about 2 to 3 feet and the part below 10 inches had become clothed with a large number of short branches. The number of lateral roots in the surface 12 inches of soil had greatly increased, especially on the second 6 inches of the taproot. The widely spreading and some of the obliquely penetrating laterals had turned downward and extended well into the second foot of soil. Finally, branching had greatly increased; older branches were somewhat longer and better rebranched and abundant new ones had arisen on the elongating roots. Thus the second and third 6-inch soil levels were rather thoroughly ramified.

MATURE PLANTS

A final study was made July 11. The well-branched vines, which were 40 inches tall, were beginning to dry at the base. The abundant pods bore peas which were half dry and too mature for use as green peas.

The pronounced taproots were traced throughout their devious course which was usually characterized by gentle curves but often by abrupt, almost right-angled turns. Depths of 3 to 3.2 feet were found. This, however, was no greater than on June 17. The number of laterals, moreover, had not increased in the surface foot of soil.

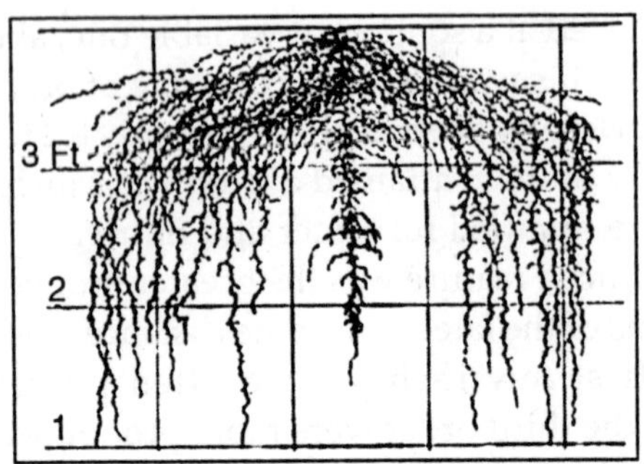

Fig. 10.2 Mature Root System of the Pea. Note the Large Soil Volume Occupied and the Greater Degree of Branching than at the Earlier Examination.

The lateral spread was a little greater (maximum, 2 feet), more long secondary laterals occurred, and branches of the third order were much more frequent. In addition to the more thorough ramification of the soil already occupied, an extensive new volume of soil had been occupied by the downwardly penetrating main lateral roots. At the June examination the longest of these did not extend beyond the second foot. At this time the entire second foot was thoroughly ramified and much of the third foot was also occupied. This added greatly to the absorbing area.

It is of interest to note that some of these roots extended to greater depths (maximum, 36 inches) than the taproot. Also, their distribution was such that the soil volume below a depth of 1 foot and to 6 inches on all sides of the taproot was almost unoccupied. Whether or not this was ramified later by an elongation of the branches of the taproot was not ascertained. At this time the taproot below the first foot was furnished with relatively short branches. Only rarely did they exceed 4 inches in length and they were frequently 0.5 to 2 inches long. The number was, as before, about five per inch but they had not only grown in length but were also much more profusely rebranched. Branches of the third order occurred on some of the older and longer laterals.

This increase in length of branchlets also characterized the larger main branches. Most of the branches of the second order ranged between 0.2 and 3 inches but not infrequently sublaterals 5 to 10 or more inches long occurred. Branchlets of the third order seldom exceeded four per inch in number and ranged from 1 millimeter to 1 inch in length. In the deeper soil, especially secondary branches often took a horizontal course. The roots were rather tough, of a tan colour, and in many cases the root ends had dried.

SUMMARY

The garden pea is characterized by a strong taproot which in its early development is profusely branched only in the first 6 inches of soil. Plants about 1.5 months old have a root depth of 2 feet. The surface soil at a depth of 2 to 8 inches is well filled with a network of nearly horizontal roots and their laterals to a distance of 18 inches on all sides of the plant. But in the deeper soil little absorbing area occurs. About a month later, when the plants are blossoming, the root system is much more extensive and efficient. The taproot is 3 feet long and much better branched throughout its entire course, secondary branches are longer and much more numerous, and branches of the third order abundant except on the youngest parts. Lateral spread has been increased to 22 inches. Many of the formerly horizontal roots have turned downward and, with those penetrating more obliquely, fairly well occupied the second foot of soil.

Nearly a month later, when the seeds are drying, the taproots have not increased in depth, nor has the number of main laterals increased in the surface soil. But the soil volume earlier delimited is much more thoroughly occupied as a result of an elongation and more profuse branching of the finer laterals. By a downward extension of the main laterals, moreover, the second foot of soil is well filled and the third foot fairly well ramified. Branches from the taproot are profuse but rather short, so that some of the deeper soil area directly beneath the plant is not fully occupied. Thus the pea completes the development of an extensive absorbing system after the beginning of blossoming.

OTHER INVESTIGATIONS ON PEAS

The taproot extended nearly perpendicularly downward to the depth of 39 inches. Below this it was too delicate to trace. Branches separated from the taproot throughout its length. These were most numerous between 4 and 8 inches in depth, where they seemed nearly to fill the soil for a distance of about 8 inches on either side. We traced a single branch root a distance of 18 inches from the taproot. The majority of the branches appeared to extend little farther than 1 foot. They gradually became shorter as the depth increased, but were 4 to 6 inches long at a depth of 30 inches. Sometimes the branches curved upward after leaving the taproot.

The American Wonder pea was also examined at the same stage of. development but the plant was only 6 inches tall. The roots extended almost exclusively downward, the taproot reaching a depth of 30 inches. No branches extended a distance greater than 4 inches from it.

Investigations in Russia indicate that although during germination and sprouting the vertical roots are rather short, later they make a uniform and rapid growth. Plants only in the second-leaf stage had a root depth of 13 inches. This was the only one of a number of vegetable crops studied that did not extend its vertical roots deeper after the inception of the flowering period. The development of the horizontal roots at the beginning was very much less than that of the vertical ones. But during the period of flowering these continued growth rapidly (after growth of the vertical roots had ceased) increasing in length from 12 to 20.5 inches. A maximum root depth of 3 feet was attained.

Numerous investigations on the pea have been made in Germany. In one it was found that it had a strong taproot that deeply penetrated with a marked early development of laterals which were especially abundant on the upper portion of the root system. These were found to extend in an obliquely outward and then downward course, many of them being nearly as long as the main root. Branching was profuse. In a good loam soil the Victoria pea was found to reach a depth of

18 inches and a spread of 12 inches when only 36 days old. Other investigations reveal a similar root habit, plants 1 foot tall having taproots 28 inches deep. The laterals attained a length of 10 inches. The upper laterals and their branches were usually much better developed than the lower ones. Occasionally, however, single branches arising deeply from the taproot made a good growth. Upon the loss of its tip, the taproot became covered with very numerous, long, much-branched laterals. These were augmented by a good adventitious root formation. Upon injury to the end of the taproot the laterals extended deeper than usual, those arising from near the injured end of the taproot growing more obliquely downward. Sometimes a single strong lateral turned vertically downward and, deeply penetrating, took the place of the taproot.

In another experiment peas of the Victoria variety were grown in filled pits in a field of well-compacted loam. The soil was washed from the roots at various stages of their development. At the age of 32 days the roots of the seedling plants were more than 10 inches deep, 56 inches at the beginning of blossoming, 69 inches at the beginning of fruiting, and about 7 feet deep when the fruits were ripe. Thus the growth in depth increased rapidly at about blossoming time and continued until maturity of the fruit. Moreover, the later root development consisted largely in an increase in root depth and not in lateral spread. Nodules occurred to a depth of 67 inches but the bulk were in the upper 6.5-inch layer of soil. Later studies confirmed the deeply rooting habit.

Another German investigator found that the pea was very similar to the common bean in its root habit. Like the bean, it formed a clearly defined taproot and then a row of adventitious roots which arose at first from the base of the hypocotyl but later, in considerable numbers, from higher parts of the base of the stem. These roots spread widely. Depths of 28 to over 31 inches were attained.89 In another study it was found that the pea was most profusely branched to a depth of 4 inches and that the largest laterals had a length of 12 inches. The number of laterals of the first order was about six per centimeter of taproot in the upper portion of the root system but only three on the deeper part. Laterals of the second order varied from two to three per centimeter. The deeper portion of the root system was only 52 per cent as well branched as the shallower part. Other European investigators have obtained similar results.

Field peas, of the variety *arvense,* were examined at Fargo, N. D., 86 days after planting and when the seed was ripe. They showed a sparse growth of roots in comparison to top development. The vines were 5.5 feet long but the roots reached a depth of only 3 feet and had rather a scanty supply of branches. The bulk of the roots was found within 8 to 10 inches of the surface.

Recent experiments with peas at Greeley, Colo., where a dwarf variety (Nott's Excelsior) with a short root system was crossed with a tall variety (Telephone) with a deep root system, indicate that root characters are hereditary and segregate out in the *F*2 generation according to the Mendelian ratios. A repetition of the experiment under somewhat different conditions of growth confirmed the results, but it was also found that the dwarf variety had almost as long a root system as the taller one. 67 This illustrates the fallacy of judging root extent by top growth. After years of study of scores of native and cultivated plants, it has been fully demonstrated that such a criterion is entirely untrustworthy.

A study of these investigations on different varieties of peas supports the conclusion that this species has a somewhat deeply rooting habit of growth. The lateral spread is also similar to that already illustrated and branching is quite profuse throughout. Considerable root growth after the time of blossoming was found by most investigators; in some varieties this growth seemed to be chiefly that of the branches, in others an extension of the taproot. Further studies, including adaptation of root system to different kinds of soil, are needed.

ROLE OF ROOT TUBERCLES

Peas and other members of the family of legumes are the

most important plants, although not the only ones, that develop nodules or root tubercles. Various strains 96 of a certain motile bacterium (*Pseudomonas radicicola*) live in the soil. They gain entrance to the root system through the root hairs and push their way back into the cortical tissue of the root where they increase rapidly and occur in great numbers. As a result of their activities, the cells of the cortex of the host make an abnormal growth which results in the well-known enlargements called *tubercles*. These frequently occur at great depths. On native legumes they have been repeatedly observed at depths of 10 to 13 feet. Although usually most abundant on roots of garden crops in the surface 8 to 16 inches of soil, they are frequently found irregularly distributed over the root system at depths of several feet.

The bacteria secure their supplies of water, carbohydrates, etc. from the host plants but build up nitrogen compounds for which the source of nitrogen is the soil air. From these compounds the legume secures large amounts of valuable nitrogenous material. When the nodules decay, due to the activities of several other kinds of bacteria, the protein contents undergo various chemical changes and are finally left in the soil as nitrates, a most favourable source of nitrogen for green plants. From 40 to over 250 pounds of nitrogen per acre may thus be added to the soil in a single season through the activities of tubercle-forming bacteria.

All parts of legumes are comparatively rich in proteins and are very valuable as fertilizers. This explains why the practice of growing leguminous crops and plowing them under has such a stimulating effect upon the growth of succeeding crops. On raw soils low in nitrates, such as railway cuts and embankments, where other plants can scarcely grow, certain leguminous plants, such as sweet clover, often thrive. On the other hand, in a soil that is too rich nodule formation is not promoted.

Experiments with the development of nodules on field peas and. other legumes have shown that they become much larger when soil temperatures are most favourable. A consistent increase in dry weight of nodules occurred as the soil temperature increased from 12 to about 24°C. At higher temperatures (about 27 to 30°C. for peas) a progressive decrease occurred. 75 Likewise nodule production decreases as soil moisture diminishes below an optimum, or may entirely cease in soils that are very dry. A soil environment favourable to the growth of roots is also favourable to the growth of nodule-forming and many other species of bacteria which promote crop growth. Thus the promotion of proper soil aeration, water content, fertility, and temperature, so far as this is possible, affects plant growth not only directly by promoting root development and activities but also indirectly through its influence upon the activities of bacteria.

ROOT DEVELOPMENT IN RELATION TO CULTURAL PRACTICE

It would seem that plants with vigorously developed and quite extensive root systems like those of the pea would thrive on many kinds of soil. In fact this has been ascertained to be true. Undoubtedly the superficial and deeper portions are modified, respectively, so as best to adapt the plant to a particular environment. Such modifications have been found for beans, the plants of which are closely related and of similar general root habit. For early crops sandy loams are preferred because they are easily warmed. But they do not retain the moisture and are often less fertile than heavier-soil types such as clay and silt loam. The roots need good aeration and the soils must be well drained. If the soil is too rich, the plants will develop large vines and ripening will be delayed.

Thorough soil preparation is especially important where the crop is sown so thickly that cultivation is not possible. This is true for a part of the market crops and the canning crop which is sown broadcast or in close drills. Poor seed-bed preparation may result in markedly decreased yields. Otherwise the root habit would indicate that beneficial results will be obtained by frequent but shallow cultivation until the roots thoroughly ramify the soil and the vines cover the ground. The practice of deep planting permits the roots to start in moist, cool soil. But it may increase the prevalence of root rot, a disease caused by a soil-borne fungus (*Thieldvia basicola*), which grows best and does greater harm when more of the stem occurs below the soil. Because of the activities within the root tubercles, peas do not require such fertile soils or as heavy applications of manure as many other crops. In fact large applications of manure do not increase the yields at the same rate as smaller ones.

PLANT BEAN

Beans belong to a family of plants (the *Leguminosae*) which, with the possible exception of the grasses, ranks first in agricultural importance. There are many genera, species, and varieties of beans. The most important beans grown in the United States are the kidney bean (*Phaseolus vulgaris*) and lima beans (*Phaseolus limensis* and *P. lunatus*). Unlike the pea, which is a hardy cool-season crop, beans are sensitive to frost and require a warm soil and a warm season for their growth.

The kidney bean is an annual, tall-twining plant. Dwarf varieties or bush beans (*P. vulgaris humilis*) are of low stature and of non-climbing habit. It is by far the most important species of bean grown in the United States. It is found in nearly every vegetable garden and is grown extensively on a commercial basis. There are 150 varieties of kidney beans in America. 66

The Wardwell's Kidney Wax variety was used for root investigations. This is a wax-podded variety of bush bean. The crop was planted May 18, an earlier planting having frozen. The plants were spaced 6 inches apart in rows 2.5 feet distant.

EARLY DEVELOPMENT

The bean seedling is characterized by a very vigorous development of the taproot. Plants only in the cotyledon stage, but grown in warm, mellow soil, often have well-branched taproots 12 inches long.

Root development in the field was first examined June 18. The plants were already 5 to 6 inches tall and had a spread of 10 inches. Each had about 10 leaves, the leaflets being 3 to 4 inches long and 2 to nearly 3 inches broad. The total leaf surface averaged 1.4 square feet. The root system was characterized by a strong, nearly vertically descending taproot and multitudes of long, horizontal branches arising mostly in the 2- to 6-inch soil level.

Some of these large branches, as in the lima bean, were adventitious, arising from the base of the stem. Although a few of the horizontal roots were over 2 feet long, their course was so devious that the maximum lateral spread was slightly less than 2 feet. Usually only about 25 roots per plant exceeded a length of 6 inches. Shorter ones were very numerous. For example, between depths of 3 and 8 inches there was usually found a total of 21 roots per inch, although at greater depths they were less numerous.

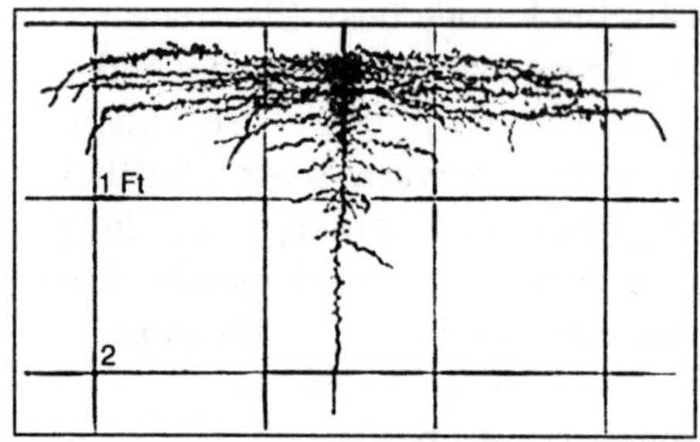

Fig. 10.3 A Month-old Root System of Wardwell's Kidney Wax Bean.

As a rule many of the shorter, horizontal roots, only 2 to 4 inches long, were unbranched or only poorly furnished with laterals. On the longer ones branching occurred at the average rate of 10 branchlets per inch. These varied from 0.2 to 2 inches in length. A very few were 4 to 8 inches long and possessed branches of the third order. Thus the soil to a depth of 8 inches was quite well occupied by a network of rootlets. These are shown in Fig. 10.4 which is a surface view of the roots in the first 6 inches of soil.

A comparison of the horizontal view with the vertical one gives an adequate picture of the root extent and position. That the roots were elongating rapidly was shown by the 3 or more inches of unbranched root ends. Also the future, obliquely downward course could already be predicted by the position of the root termini.

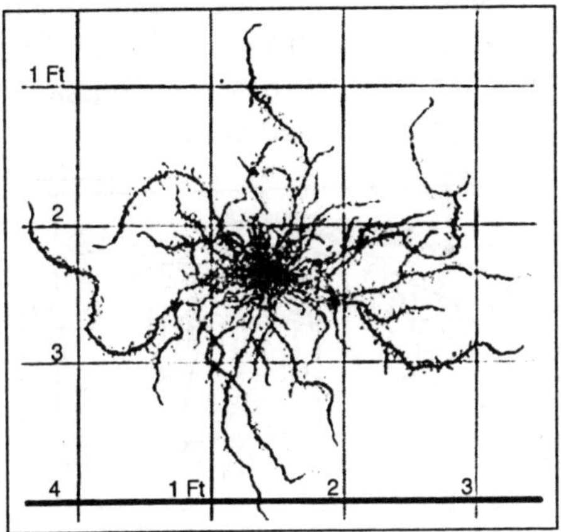

Fig. 10.4 Surface View of the Roots of the Bean in the First 6 Inches of Soil.

The taproot, below the 8-inch level, was well clothed with rootlets (Fig.10.5). Only those exceeding 3 to 6 inches in length were rebranched. The last 5 inches of the turgid, thick, white taproots were quite smooth, although considerably kinked and curved from penetrating the compact soil. A maximum depth of 27 inches was ascertained.

MIDSUMMER GROWTH

The plants were now nearly 1 foot high and had a spread of approximately 1 foot. Those of average size possessed about 18 large leaves, the largest of which were over 1 foot in width. The largest leaflets had a length and diameter of 6 and 5 inches respectively. The transpiring surface had increased to 5.5 square feet. The plants were blooming profusely and numerous young fruits were growing vigorously.

The roots too had made considerable growth, although this was not so marked as in many other vegetable crops. The lateral spread reached a maximum of 30 inches, the most widely spreading roots sometimes ending in the surface foot of soil.

Where the tips of the roots had been injured or destroyed, numerous laterals arose from the root ends and some of them pursued the course which would have been taken by the main root. This phenomenon of the growth of laterals being promoted by injury to the main branch has been repeatedly observed among many vegetable and field crops.

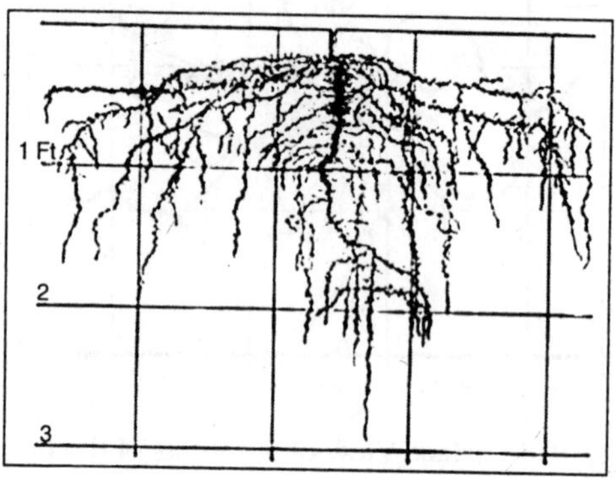

Fig. 10.5 Root System of Kidney Bean about 2 Months Old.

The taproots had also increased their depth from about 27 inches at the June examination to 3 feet. Likewise the working level (exclusive of taproot) which scarcely exceeded 12 inches, had now been extended to an approximate depth of 20 inches.

As before, considerable variation was found in the total number of large main roots. On the plant selected for drawing, most of these, as in the lima bean, arose from the base of the enlarged stem. Sometimes the taproot furnished more of them. A comparison of Figs. 10.4 and 10.5 shows that the roots on the plant selected to represent the later stage are scarcely as horizontal as those in the earlier drawing. This is largely due to variations in the individuals. The downward growth of these widely spreading, as well as the more obliquely growing, roots is here well under way.

These strong main or lateral roots had diameters of 2 to 3 millimeters and were at least 1 millimeter thick throughout their course. The rate of branching was, as before, about 10 laterals per inch of root. They varied from 3 to 15 but usually approached the higher number. Many of them were only 0.2 to 1 inch in length, especially on the younger portions of the older roots and on the longer branches of the old ones. The latter were now abundant and ranged from over 1 inch to 6 inches in length. They were usually profusely rebranched, roots of the fourth order being not uncommon. Frequently, the branches occurred in clusters of threes on opposite sides of the roots. The general direction of these large branches, their lax and often sinuous course, and the degree of branching may best be visualized by a study of the drawing.

As regards the taproot, it plays an important rôle. Filling the soil with its numerous branches just below the plant and extending considerably beyond the working level of the other roots, it continually drew upon new sources of supplies. Although some taproots were quite vertical in direction of growth, others pursued a very devious route as is shown in the drawing. As before, many short branches came from the older, shallower portion. These were better rebranched than at the earlier examination. Moreover, numerous, long, well-branched laterals, running in various directions, had developed and added greatly to the absorbing area. As on all the major roots, and indeed many of the smaller ones, the 3 or more inches of unbranched, glistening white tips showed that the plants were still growing rapidly.

MATURE PLANTS

A final study of the kidney bean was made Aug. 5. The very leafy plants were about 13 inches tall and had a total spread of 15 inches. Although some were still blossoming, most of them had an abundance of mature pods. That they were fully grown was shown by the drying of some of the leaves.

The roots had not increased in lateral spread but the taproots and some of the longest main laterals (including those arising adventitiously from the stem) had extended to depths of 40 to 46 inches, The working level had also been greatly increased, from 20 to 36 inches. Root branching was very well developed to the 3-foot level and to a distance of 2 feet on all sides of the plant. Many roots extended deeper. A few, especially the deeper portion of the taproots, showed signs of decay.

SUMMARY

The kidney bean rapidly develops a deeply penetrating taproot. On plants 31 days old and 6 inches tall this reaches a length of over 2 feet. Branching is profuse throughout but especially in the 10 inches next to the surface. Roots from the base of the stem and from the taproot extend horizontally but deviously in the second to the eighth inch of soil to distances of 12 to 24 inches. With the numerous branches they constitute the bulk of the absorbing system which is distinctly superficial. A month later, when blossoming is profuse and fruits beginning to form, the roots are more extensive. The taproot has increased its depth to 3 feet and the working level to 20 inches. It is widely branched even in the second foot of soil. The main horizontal branches have extended the lateral spread to 30 inches and have branches which penetrate far into the second foot of soil. When the plants are nearly mature, the soil to 2 feet on all sides is well ramified to a working level of 3 feet and numerous roots extend 1 foot deeper.

Thus the general root habit of the kidney bean is not greatly unlike that of the pea although the lateral spread and depth of penetration are somewhat greater and the deeper soil, just beneath the plant, somewhat more thoroughly occupied.

LIMA BEAN

The large lima bean (*Phaseolus limensis*), although a perennial in the South where it is of considerable commercial importance, is grown as an annual in northern United States. It is a vigorous grower but requires higher temperatures and a longer growing season than the common garden beans. It is a stout, high-climbing plant although bush or dwarf forms do not climb.

Burpee's Bush Lima bean (variety *limenanus*) of the large-seeded, flat type was planted June 2 in rows 30 inches distant. The seeds were placed 6 to 12 inches apart in the row.

EARLY GROWTH

The root system was first examined 2 weeks later on June 18. The cotyledons were yellowing and nearly every plant was furnished with two leaves each about 4 inches long and one-half as wide. The total leaf surface was 24 square inches. The plants averaged 5 inches tall.

A strong taproot extended from the base of the stem and reached a depth of 13 inches. The very numerous, mostly horizontal branches were longest above (maximum, 9 inches) and gradually became shorter at greater depths, thus giving a conical shape to the root system as a whole. No roots were found in the surface 2 inches of soil but numerous, large, adventitious roots arose from the base of the stem and spread widely just below this soil level. They were growing rapidly and the last 3 inches were quite unbranched. Otherwise branches of the first order only 0.3 to 2 inches long (but mostly short) occurred at the rate of about 8 per inch. From the older portion of the taproot, the first 5 inches, laterals arose in great numbers, often as many as 30 per inch. These varied in length from about 4 inches on the older part of the taproot to 1 millimeter on the younger portion. The last 4 inches of the taproot were unbranched.

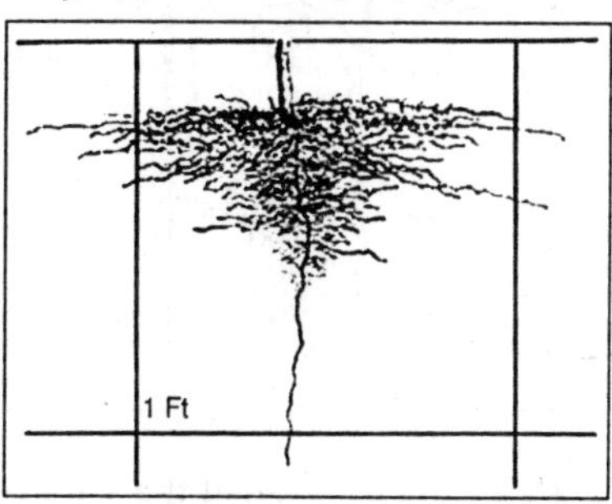

Fig. 10.5 A 16-day-old Root System of Burpee's Bush Lima Bean.

MIDSUMMER GROWTH

The plants were 10 inches high and in bloom. The stout

stems bore approximately 33 leaves each. The largest leaflets were 3.5 by 2 inches in length and breadth, respectively, and the total transpiring surface was 4.3 square feet.

The root system also had made a good growth. The lateral spread had increased to over 2 feet and a depth of nearly 3 feet had been attained. A root system was usually composed of a taproot and 6 to 8 large laterals arising from the base of the enlarged stem, although there always were numerous, smaller rootlets. The taproot pursued a somewhat devious downward course, often kinking and turning either in long, graceful curves or abruptly and penetrated deepest, 33 to 35 inches. The more superficial large roots spread laterally, often 1 to 2 feet in the surface 8 inches of soil, and then turning more obliquely or even vertically downward reached depths of 20 to 30 inches. Still other main roots, running more obliquely downward, filled in the soil volume between the widely spreading laterals and the area occupied by the taproot and its branches. These large roots were 1 to 2 millimeters in diameter and maintained this thickness throughout their course.

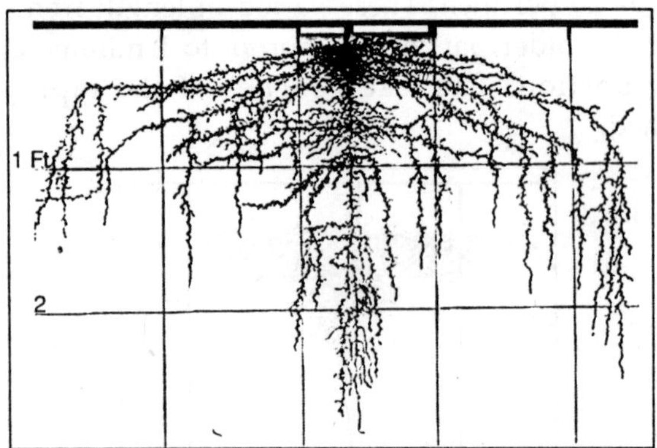

Fig. 10.6 Root System of Lima Bean at the Time of Blossoming.

Branching on the laterals was somewhat variable but, in general, it was profuse. The shorter branches were simple, mostly only 0.3 to 2 inches long, and occurred usually at the rate of 4 to 12 per inch. The longer branches, *i.e.*, those exceeding

2 to 3 inches in length were practically all furnished with laterals and at a rate similar to that of the main branch. Frequently, they were 9 to 15 inches in length.

On the taproot branching was very profuse; 25 to 28 laterals 0.2 to 4 inches long occurred on each inch through the first few inches of its course. Branching was especially pronounced in the furrow slice (first 8 to 9 inches, Fig. 10.6), some of the horizontal branches spreading rather widely and sometimes giving rise to long, vertically penetrating rootlets. At greater depths they showed a distinct tendency to spread a few inches and then turn downward, often paralleling the course of the taproot for 1 foot or more. The longest branches were already furnished with tertiary roots. Thus the plants had a very efficient absorbing system.

MATURING PLANTS

The final examination was made Aug. 25. The plants were well developed and in fruit. They were 19 inches high, had a total spread of tops of 30 inches, and had not yet begun to dry. About 25 branches were found on plants of average development. These were covered with an abundance, of large, green leaves Many large pods, nearly 5 inches long and over 1 inch wide, as well as numerous smaller ones, were found on normally developed plants. Blossoming had just ceased.

The underground parts had made a striking development. The lateral spread had doubled, now extending to a maximum of 4 feet; the depth of penetration had been increased from 3 to 5.5 feet; and the working level to 45 inches. Usually the stem extended 2 to 2.5 inches into the soil and gave rise to a deep taproot on its terminal portion and almost equally large laterals from near its base. The latter were usually 7 or 8 in number and in length and degree of branching they were often quite as important as the taproot. In cases where these lateral roots were fewer (sometimes only 4 in number) the taproot gave rise to many strong, well-developed laterals. The taproot and the large adventitious roots from the stem varied from 1.5 to 3 millimeters in thickness, usually the taproot having the larger size. Nearly all maintained this diameter for the first 12 to 18 inches of their course and some maintained it through a distance of 5 feet or more to their ends. Occasionally, a root would taper to 1 millimeter in thickness and then again enlarge.

The taproot alone pursued a generally vertically downward, although somewhat meandering, course and extended more deeply. Most of the other main roots ran 1 to 3 or more feet in the surface 6 to 12 inches before turning downward.

They sometimes forked, however, and gave rise to groups of three to five branches which ran more or less vertically downward into the third or fourth foot of soil. Moreover, nearly all of these roots gave rise in their horizontal course to vertically descending, major branches. Some of these grew only 6 inches from the taproot, others grew over 3 feet from it. These roots, with-their branches very completely occupied the soil to the working level at 45 inches.

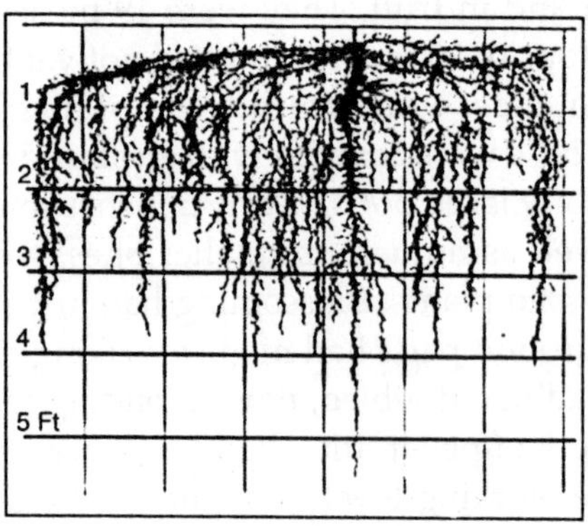

Fig.10.7 The Great Extent of a Maturing Lima Bean Plant is here Shown. Two Hundred Cubic Feet of Soil were Ramified by the Roots of a Single Plant.

The rate of branching was somewhat variable, both as to number and length of laterals, but it was always profuse. A rate of 6 to 12 per inch was common, and 13 to 16 per inch not unusual. The smaller branches ranged from 0.1 to 5 inches in length. Most of these, especially the longer ones above the 2.5-

foot level, were supplied with laterals of the second order at a similar rate. Rootlets of the third order were not uncommon. Longer laterals, many of which have already been mentioned as descending vertically, were rebranched like the main roots. Although the first few inches on the main lateral roots were not so well furnished with branches, this area was extremely well occupied by short laterals (0.5 to 2 inches long) arising from the taproot. For example, 72 of these were found between depths of 3 and 9 inches. These thread-like outgrowths were usually unbranched. Between the depths of 8 and 15 inches the roots were only about half as numerous, usually I to 4 inches long with 10 to 15 branches per inch. These were rebranched often to the second order, thus forming an efficient root network surrounding the taproot. Below this to a depth of over 2 feet branching was again more abundant and the branches averaged much longer. They were very short and little absorption occurred below the 4-foot level.

SUMMARY

The root system of the lima bean is very similar to that of the kidney bean. In early growth it develops a taproot with very abundant horizontal lateral branches; others arise adventitiously from the base of the stem. The root, system is conical in shape, the base of the inverted cone beginning, about 2 inches below the soil surface. During the 5 weeks following, about eight strong lateral roots spread 2 feet or more in the surface foot of soil, branching profusely. Turning downward they end in the second or third foot. Large numbers of smaller laterals fill the soil near the taproot. The taproot is also covered with long laterals almost to its end, approximately at the 3-foot level, which greatly increases its absorbing area.

By the end of blossoming the root system has made further marked growth. Widely spreading, nearly horizontal branches run outward 2 to 4 feet in the surface 12 inches. Then they often turn downward and, like the long laterals arising in their horizontal course, penetrate deeply, frequently far into the fourth foot of soil. More obliquely descending long roots and shorter branches arising from the taproot (which was 5.5 feet long) fill in the remaining soil volume just beneath the plant. Often the major branches are as important as the taproot. Where they are few, the deeper portions of the taproot are better branched. An intricate network of small rootlets occurs throughout the 200 to 225 cubic feet of soil ramified by a single plant. The lateral spread of roots is about 4 feet. This is 1.5 feet greater than that of the kidney bean, and the working depth is about 9 inches deeper. The larger root system of the lima bean correlates with the greater development of tops.

OTHER INVESTIGATIONS ON BEANS

Many studies have been made upon the root habits of beans especially in Europe. Certain German investigators have found that the bean *(Phaseolus vulgaris)* and the vetch or broad bean (*Vicia faba*) are very similar in their root habits. Both of them form a clearly defined taproot, and then rows of adventitious roots. The latter arise at first from the base of the hypocotyl but later, with the increasing age of the plant, they arise in considerable numbers from higher parts of the stem base. In the bean, these were found to be developed almost as strongly as the main root. The stronger root branches of the first order, which occurred in great numbers, especially in the bean, penetrated to about the same depth as the main root. The depth of penetration in the bean was 40 to 43 inches. The lateral spread varied from 20 to 30 inches. 89 Similar results in regard to depths of penetration have been found by other investigators.

Two varieties of *Phaseolus,* Pride of Lyon and Princess of Orleans, have been carefully studied in Russia. The first had the same depth of root penetration (25 inches) at the age of 45 days as did the latter when it was 10 days older. At the beginning of the flowering period the relative depth and lateral spread for the two species were 26 and 10 inches for the former and 30 and 12 inches for the latter. The fact that different varieties have a different root extent was clearly indicated. Princess of Orleans had a weaker growth of roots which ceased development earlier than the more extensive root system of the

Pride of Lyon. The final root depths were about 34 and 38 inches and the lateral spread was 12 and 20.5 inches, respectively. Beginning with the flowering period, the horizontal roots increased in length only slightly.

The Boston Dwarf Wax bean was grown at Geneva, N. Y., and washed from the soil about the middle of August when the plant was approaching maturity. The soil was a clay loam, 6 to 10 inches deep, underlaid by a tenacious gravelly clay subsoil. The deeper roots were traced to a depth of 2 feet and the horizontal ones extended quite as far on either side of the plant. The Scarlet Runner bean was also examined at the same time. The stems were about 4 feet high and the plant was in full bloom. The deeper roots extended to 2.5 feet and the longer horizontal ones at least a distance of 4 feet. A few roots grew within 1 inch of the surface, but the great majority were between 2 and 8 inches in depth. The root system was very similar to that of the wax bean but decidedly more extensive.

Other studies have shown that the kidney bean, the scarlet runner, and the lima bean are all characterized by a rather strong taproot. When the taproot meets obstacles, it is easily turned aside from its course or may break up into many branches. Laterals of the first order are very numerous, develop early, and axe abundantly clothed with much-branched laterals of the second order. The lateral spread was found to be greater than that of the pea. Bean plants only 36 days old reached a depth of 12 inches and had a lateral spread of 20 inches.

It has been found that the root system of the kidney bean is capable of considerable modification. In a loose, rich soil the roots are able to extend their depth of penetration by promoting the growth of the taproot and strong, basal, outwardly and downwardly penetrating roots. This is accompanied by a decrease in branch production, somewhat directly proportional to depth. But if the root tips are disturbed by cultivation, the laterals arising from these roots, which are already long and thick, are much more limited in downward growth. Thus the bean also has the ability to adapt itself to shallow soil. Since the stronger roots are hindered from penetrating deeply, they produce more numerous and longer laterals and the basal roots arising later pursue a more horizontal course.

The broad bean (*Vicia faba*) has a root habit very much like that of the navy bean with strong, widely spreading laterals arising from the upper portion of the taproot and more obliquely penetrating branches from the deeper part. It was experimentally determined, by growing beans in boxes 6 and 14 in. in depth, respectively, and filled with the same volume of soil, that the taproots were much shorter in shallow than in deep soil, but that the number of secondary roots per unit length was greater in the shallow soil. When the taproot was cut, no injury to the plant ensued provided the cutting was far enough below the surface so that sufficient taproot was left to produce an abundance of laterals.

Free development of the taproot to a considerable depth was found unnecessary to the best growth of the plants under the conditions of the experiment. But where the deeper layers of the soil were to be utilized, the taproot is needed to give rise to the deeper laterals. That the plant adapts itself to shallow soil by modifying the usual root habit is a fact of great significance and of much importance to agriculture.

In another experiment broad beans were grown in a field of loam soil. When the plants were 4 to 5 inches tall, the upper lateral roots were cut close to the plants to a depth of 6 inches. This was done at midday early in June. The weather was clear and hot, following ample rains. The plants soon wilted and, in fact, wilting was noticeable for several days. On some of the plants the lower leaves died. The plants grew poorly and had increased very little in height by July 1 when a second, similar cutting was made.

They did not wilt following the second cutting, but even a month later they were much more poorly developed than uninjured plants. On the other hand, cutting of the taproot of other plants below the main branches, *i.e.,* at a depth of 6 inches, did not result in wilting although at the time this was done the soil was quite dry. It was demonstrated by other experiments that deep, tillage gives greater returns where the plants are widely spaced than where they are grown closely together. The relation to root injury is apparent. Other investigations on the broad bean showed that it was most profusely branched to a depth of 4 inches and that the largest laterals had a length of 10.5 inches.

The number of laterals of the first order was about seven per centimeter of taproot in the upper portion of the root system but on the lower part there were only about two per centimeter. The deeper portion of the root system was only 29 per cent as well branched as the shallower part. Thus, in comparison with the pea examined at the same time, the maximum length of branches was 1.1 inches less, but the number of primary laterals was somewhat greater in the surface soil. On the primary and the secondary laterals of the pea the branches were two to three times as numerous in the deeper soil and the total branching 23 per cent greater (relative to the shallower portion of the root system) than in the case of the bean.

The roots of the large, flat lima bean, grown in California, extend to depths of 3.5 to 6 feet. The taproot penetrates almost. straight downward to this depth, giving rise to smaller and more fibrous rootlets at intervals. Where the water table is at such a level that the roots of the plants penetrate down to the moisture which arises from it by capillarity, the beans do exceedingly well and are not so badly affected by the hot, dry days. They will endure very dry soil on the surface after their roots become established below. In tropical climates the root is sometimes large and fleshy and the plant lives more than 1 year. A survey of these investigations on the rooting habits of beans points clearly to the fallacy of the current idea, commonly expressed in the literature on vegetable growing, that beans are a shallow-rooted crop. It is true that a portion of the root system lies just beneath the surface of the soil, but the roots also extend widely and penetrate deeply.

Among the varieties studied considerable differences were found in the rooting habit and the ability of the root system to adjust itself to different soil environments has been shown. Further studies of these and others of the very numerous varieties will throw much light upon root behaviour under various conditions of growth. How the roots respond in their distribution and activities under dry-land culture, irrigation, various methods of seed-bed preparation, depth and method of distributing manures and other fertilizers, depths of intertillage, etc. are problems awaiting further investigation. 85 More attention should be given to their activities in the subsoil. These will be found not only of great scientific interest but may lead to modifications and improvements of current cultural practices.

ROOT HABIT IN RELATION TO CULTURAL PRACTICE

The widely spreading roots of beans, the fact that they are richly furnished with tubercles, and the plasticity of the root system in adapting itself to various environments, all lead to the conclusion that beans will grow well in various types of soil. The fact that they do quite well in practically all types from light sandy loams to heavy clays is well known by most gardeners. That the common beans will thrive on poor soils better than many other crops is expressed in the vernacular, "It is good enough for beans." In the North, however, clay soils may be too wet and warm too slowly, and sandy soils may be too droughty.

The plants respond in root development and yield when grown in moderately fertile, carefully prepared, well-drained but moist and properly cultivated, non-acid soil. The bean plant is especially sensitive to poor tilth and rough treatment at the time of cultivation. A good seed bed, well pulverized and firmly packed, permits the maximum utilization of the soil nutrients and moisture throughout the entire growing season. In a poorly prepared seed bed the fibrous roots may die and the plant be handicapped by the partial or complete loss of that portion of its root system occupying the richest part of the soil. On soils that are too rich there is a tendency towards too great a growth of vines and, undoubtedly, a compact root system. Since both kidney and lima beans are epigeal in habit and the large cotyledons must be pulled out of the ground, thorough preparation and good tilth of a well-compacted seed bed are essential. A deep, mellow soil promotes deeper root development and at the same time a more thorough occupancy of the richest portion of the substratum, the surface soil layer. This is due in part to better aeration and partly to better water-holding capacity of the soil. In dry-land farming this early stirring of the soil forms a dry soil layer. If this is maintained, the plants will root only below this layer. Otherwise many surface roots would undoubtedly die as a result of later drought.

Under irrigation, where a furrow is to be made between the rows to permit watering while the crop is growing, it is the practice to cultivate deeply this part of the soil. This is a precaution to keep the soil dry and thus prevent the roots from extending into the area so that they will not be injured when the furrow is opened. 114Hoeing the soil about the stems promotes the production of very many adventitious roots. Subsoiling to a depth of 16 to 18 inches is advocated to increase moisture penetration in more and districts. 55 In certain sections where beans have been grown for many years, it is necessary to practice crop rotation as a preventive measure against dry root rot. This disease is caused by a soil-borne fungus (*Fusarium martii phaseoli*). It appears as a reddish-brown dry rot at the base of the plant and often causes the loss of almost the entire portion of the finer roots as well as the shriveling of the end of the taproot.

Early cultivation close to the plant is undoubtedly beneficial but the rate of lateral root growth should be carefully considered. By adapting their roots to a shallow soil, beans are limited in their range for securing nutrients. Their roots are thus unfavourably situated for enduring drought. The plants quickly develop and soon completely cover the ground, preventing rapid evapora tion of moisture from the soil should it escape absorption by the network of surface rootlets. Early cultivation should be shallow to avoid root injury. A late crop of beans showed very little advantage in cultivation over keeping the weeds down by scraping the surface. Proper spacing of the plants is an important feature of vegetable production. Crowding reduces the growth of tops and this in turn results in a poorly developed root system. Larger varieties and plants grown on rich soil where there is little danger of drought or where irrigation is used are more widely spaced. Where moisture is the limiting factor in growth, as in dry farming, wide spacing is imperative. For example, in Colorado under irrigation the rows are spaced 28 inches distant, but on dry lands they are spaced 36 to 42 inches apart. The plants are spaced 4 to 6 inches apart in the row under the first condition and 10 to 12 inches under the latter.

Drilling the seeds 4 to 6 inches apart gives a better distribution of the plants than placing five to eight seeds in a hill 2.5 to 3 feet distant. This former method at least partially eliminates crowding. Experiments have shown that where field beans were planted at the rate of five per hill, and the hills spaced 18 inches apart in rows 3 feet distant, the yield was hardly more than one-half of the yield where the beans were spaced 6 inches apart in the row. 28 Garden beans are usually spaced too closely 38 as may be readily seen by an examination of the rooting habit.

OKRA

Okra or gumbo (*Hibiscus esculentus*) is a stout, erect, branching plant, 1.5 to 6 feet tall. Like cotton, it belongs to the family of mallows. The plants are cultivated as annuals for their large, soft, immature pods. Although of no great importance in the United States, it is commonly found in home gardens in the South. It is a tender plant and thrives best in hot weather.

Seed of the Mammoth long-podded variety was planted at Norman, Okla., in rows 3.5 feet apart. Several seeds were placed in a hill, but later the seedlings were thinned to single strong plants 3 feet apart.

EARLY DEVELOPMENT

About 3 weeks later the plants were 5 inches tall and each possessed 4 leaves with blades 2 to 2.5 inches in length. Okra has a strong taproot which penetrates almost vertically downward. The taproot was about 5 millimeters in diameter and reached a depth of 16 inches. A total of 24 to 35 laterals, the largest 1.5 millimeters thick, ran horizontally from just beneath the soil surface to a depth of 8 inches. The roots originated on four sides of the taproot. A maximum spread of 18 inches was reached at the 5-inch level. A few of the deeper laterals pursued an obliquely downward course. All were furnished with a profuse growth of laterals 0.1 to 3 inches in length. The rate of branching was about 8 laterals per inch (Fig. 10.8). The roots were white and rather tender.

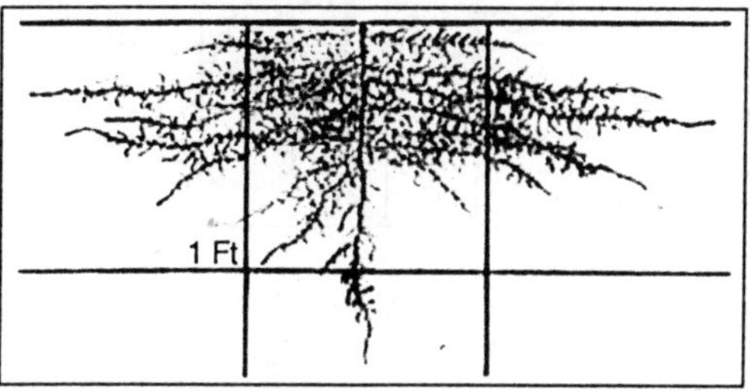

Fig. 10.8 One-half of the Root System of Mammoth Long-podded Okra about 3 Weeks After Planting.

LATER DEVELOPMENT

The branched stems had reached a height of 8 inches and were ½ inch in diameter and clothed with numerous leaves, the largest being 8 inches in width. Flower buds were appearing. The strong taproots tapered rapidly from 12 millimeters in diameter near the soil surface to 1 millimeter at a depth of 10 inches. They pursued their rather vertically downward course to depths of 20 to 22 inches. The strong laterals in the surface 8 inches of soil ran in a generally horizontal course to distances of 15 to 32 inches from the base of the plant. A total of 29 roots was found on a plant of average size on the first 8 inches of taproot. These are shown in surface view in Fig. 10.8. The widely divergent roots were branched at the rate of 4 to 6 laterals per inch; only the longest bore sublaterals. Thus a rather large volume of the rich, moist, surface soil was quite well ramified. Below 8 inches the branches were almost as numerous but decreased rapidly in length towards the growing tip from a maximum length of only 5 inches. The white, slightly fleshy roots were rather tough.

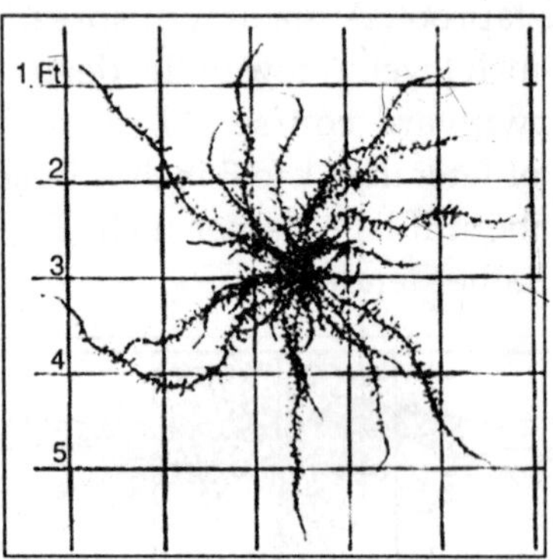

Fig. 10.9 Surface View of the Roots of Okra Occupying the First Foot of Soil.

MATURING PLANTS

By July 22 the well-branched plants were 4.5 feet tall and had a stem diameter of 2 inches. They had made such a vigorous growth that the soil between the 3.5-foot rows was quite concealed. The plants had been fruiting for some time and the pods were exceptionally large. Growth, however, was proceeding vigorously.

The taproot, now nearly 2 inches thick near the soil surface, gave rise to so many major branches that it tapered rapidly, and was only 3 millimeters in diameter at a depth of 1 foot. It pursued a rather devious downward course to a depth of 4.5 feet. The laterals of the surface 8 inches had increased enormously in size. From 11 to 17 were found on various plants with diameters of 5 to 20 millimeters. Even 2 feet from the base of the plant some of these strong roots were 3 millimeters thick.

The maximum lateral spread had been increased from 32 inches to 79 inches. Many roots, of course, did not spread so widely. But nearly all of those originating in the surface 8 inches ended near the soil surface or turned downward near their extremities for only a few inches. The branches below the 8-inch level had also made a vigorous development. Several laterals 1 to 2 millimeters thick originated from the taproot at depths of 10 to 20 inches.

These ran outward and downward or outward for only a few inches and then turned abruptly downward and paralleled the course of the taproot. Frequently, they reached depths even greater than that of the taproot. Aside from these larger roots numerous others occurred in the four rows on the sides of the taproot. Usually about 8 to 12 roots arose from each inch of the taproot.

Large branches, sometimes 3 to 4 feet long, now arose from the major branches. In general, these pursued a horizontal direction. Like the other roots they were clothed with a great network of rootlets at a rate varying from 4 to 12 per inch,

Young plants of okra have a strong taproot and many nearly equally long, mostly horizontal branches whose numerous laterals fill the surface 8 inches of soil. When the flower buds appear, the 25 or more horizontally spreading roots extend from 0.5 to 2.5 feet from the base of the plant. They are well furnished with absorbing rootlets. The taproot attains approximately a depth of 2 feet, but below 8 inches the branches are short. Mature plants have taproots 2 inches thick and 4.5 feet deep. The surface laterals become much thickened and extend widely, some to 6 feet. Some turn downward a few inches near their ends but none penetrate deeply.

Several branches originate below the 8 inch level; running obliquely outward and then far downward, they supplement the absorbing area of the well-branched taproot. Thus the root system consists of two rather distinct parts: a shallow, widely spreading portion thoroughly ramifying the surface 18 inches of soil to a distance of 6 feet on all sides of the plant; and a deeply penetrating, well-branched taproot which gains access to the water and nutrients in a soil column about 2 feet thick and which extends to a 4-foot level. This latter portion of the root system develop rather late. Thus the soil between the widely spaced plants is thoroughly occupied.

CARROT

The carrot (*Daucus carota sativa*) is either an annual or biennial plant. Early varieties seed the same growing season in which they are planted. Later ones produce only the whorl or large cluster of fern-like leaves the first year and extend the rough, branching flower stalks to a height of 2 to 3 feet during the second season of growth. This fairly hardy, rather slowly maturing vegetable is common in most home gardens as well as in many market gardens. The "carrot" mostly consists of enlarged taproot but the upper portion develops from the hypocotyl and is a part of the stem.

Seed of the Chantenay variety of carrot was planted Apr. 24 in drill rows 18 inches apart. The seedlings were thinned to a distance of 5 inches in the row.

EARLY DEVELOPMENT

The first examination of root development was made June 10. The tops were 4 inches tall and consisted of five leaves each, the larger ones had blades 2 by 2 inches in outline and the average transpiring area, as determined by the aid of a planimeter, was only 12.5 square inches.

The plants were characterized by strong taproots 4 to 6 millimeters in diameter. These soon tapered to 1 millimeter in thickness, a diameter held throughout their vertically downward course. Depths of 27 to 32 inches were attained. Compared with other garden plants, the taproots were poorly branched (Fig. 10.10). No branches occurred in the surface inch of soil but to a depth of 6 inches they arose at the rate of five to six per inch In the deeper soil branches were fewer and unevenly distributed. They varied in number from one to nine (average, three) per inch. Nearly all ran in a rather horizontal direction. They varied in length from 0.2 to 9 inches, the longest Qnes always occurring on the oldest portion of the taproot. Secondary branches, even on the oldest laterals, were not abundant. The last 6 to 9 inches of the taproot were entirely free of laterals. The poor branching and consequently relatively small absorbing area may be correlated with the small transpiring surface afforded by the deeply incised leaves. Even the root-hair development was much less pronounced than on many other garden plants. The roots are very brittle.

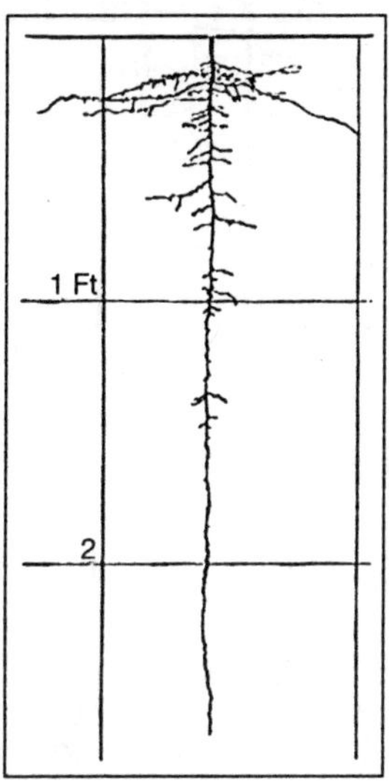

Fig. 10.10 Taproot System of a Chantenay Carrot.

MIDSUMMER GROWTH

A month later, July 12, the plants had reached a height of 13 inches and attained a spread of 16 inches. The shoot had a thickness of ¾ inch near the base. Each plant had about 14 leaves. Many of these had blades 5 to 6 inches wide and 12 to 14 inches long. Thus the transpiring and photosynthetic area had greatly increased.

The fleshy portion of the taproot extended to a depth of 6 to 8 inches. It had a maximum diameter of slightly more than 1 inch and tapered gradually at 8 inches depth to a thickness of about 2 millimeters. In the deeper soil (below 18 inches) the taproot was only 1 millimeter thick. Its course through the deeper soil was quite tortuous. Abrupt turns and deviations of 0.5 to 2 inches from the vertical were frequent (Fig. 10.11). Depths of 4 to 4.8 feet were usual.

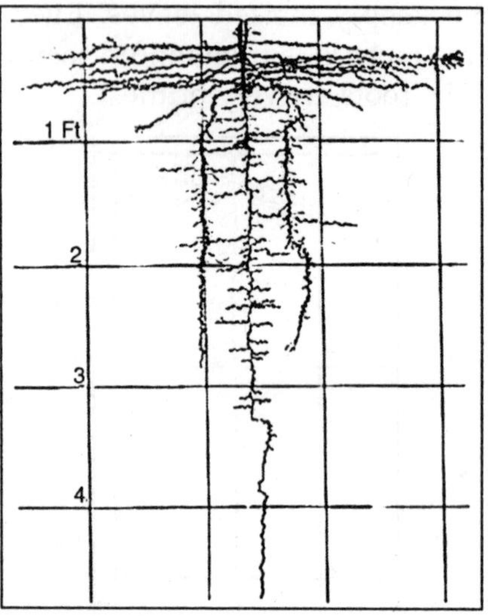

Fig. 10.11 Root System of a Carrot

Only a few, short branches arose in the surface 2 inches of soil. These were unbranched. Between 2 and 9 inches in depth, however, 45 to 55 laterals originated. Usually 16 to 24 of these spread horizontally for distances of 15 to 24 inches. A few of those which originated deepest (5 to 9 inches) turned downward and frequently extended to the 2- to 2.7-foot level. Although the other branches in this region of the taproot were shorter, thread-like, and poorly branched, these longer roots (about 1 millimeter in diameter) were clothed with laterals at the rate of 5 to 9, but sometimes 10 to 15, per inch. They were 4 millimeters to 2 inches (rarely more) in length and most of them were unbranched. A few of the main laterals that had been injured terminated in brush-like mats of rootlets. Below the 9-inch level roots arose at the rate of three to four per inch throughout the course of the taproot except on the last 6 to 9 inches which were devoid of branches. Below 3.5 feet these were short and unbranched. Nearly all pursued a rather horizontal course. Many short, unbranched or meagerly branched roots alternated with a few of greater length which were very well clothed with rootlets. The relative lengths, degree of branching, etc. are clearly shown in Fig. 10.11 where a carefully selected root is pictured. The older roots vary in colour from tan to yellow. Only the younger portions are white.

MATURING PLANTS

At the final examination, Aug. 12, plants of average size had 16 fully grown leaves and 4 to 6 partly developed ones. Some of the older leaves had considerably deteriorated. The fleshy portion of the taproot was about 8 inches long and 1.5 inches in diameter near the soil surface but less than ½ inch thick at the 8-inch level, the Chantenay being a half-long, stump-rooted variety. The average number of branches at the several depths is shown in Table.

Table. Number of Laterals Originating from the Taproot of Carrot

Depth Inches	Small Branches	Large Branches	Depth, Inches	Small Branches	Large Branches
0-1	7	0	14-15	5	0
1-2	9	1	15-16	4	0
2-3	13	0	16-17	7	0
3-4	5	0	17-18	7	1
4-5	6	2	18-19	4	1
5-6	5	3	19-20	2	0
6-7	7	1	20-21	6	3
7-8	5	0	21-22	11	0
8-9	4	0	22-23	10	1
9-10	3	1	23-24	6	1
10-11	1	1	24-25	6	2
11-12	1	1	25-26	3	0
12-13	11	0	26-27	7	1
13-14	7	1			

Thus on the first 2.3 feet of a taproot of average size approximately 20 large roots and 160 small ones had their origin. The large ones varied from 1 to over 2 millimeters in diameter. A few near the surface ran outward 12 to 18 inches and, like the network of shorter, smaller roots originating from the fleshy portion of the carrot, branched profusely and ended in the surface soil. But many of the larger roots, after pursuing a horizontal course for 1 to 2 feet (maximum spread, 2.3 feet) turned downward. They then ran almost straight downward ending in the 3- to 6-foot level. A comparison of Figs shows the characteristic development of these roots. At the earlier stage of growth they had not reached their maximum spread. Moreover, only a few were utilizing the moisture and nutrients of the deeper soil.

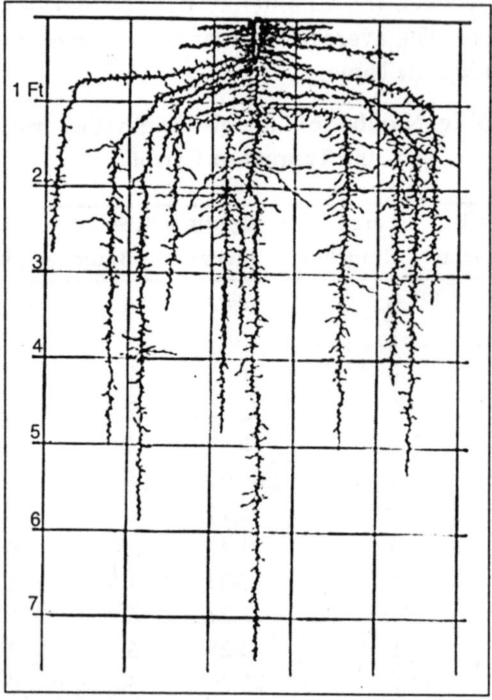

Fig. 10.12 Root System of Carrot.

Secondary branches on all of the roots were not only more abundant (4 to 13 per inch) but also longer (now 2 to 3 inches) than at the previous examination. Considerable variation occurred in the distribution of laterals. Some were 6 inches in length. Branches of the third order were nowhere abundant. An examination of the drawing shows that many new roots had arisen from the deeper portions of the taproot where they averaged 7 per inch. The average length, however, was only slightly greater than at the previous examination. The taproots were traced to their ends at depths of 6 to 7.5 feet. The roots were still growing vigorously. Judging from the development of such crops as parsnips and beets excavated in September, it seems certain that the roots of the carrot reach a working level of 6 to 7 feet and a maximum depth at least of 10 feet.

GROWTH OF CARROTS DURING WINTER AND SPRING

Well-developed carrots of the Long Orange variety, with the fleshy taproots 2 inches in diameter, were left in the soil at Norman, Okla., at the end of the 1925 growing season. Because of late-summer drought there were only a few green leaves during August and September. Death of all or nearly all of the absorbing roots in the surface foot of soil resulted from the drought. Practically no growth either above or below ground occurred from the middle of August until after the September rains. By Dec. 3 there was a vigorous growth of leaves to a height of 8 inches and the roots also had made renewed growth. From the same areas on four sides of the fleshy taproot that had previously given rise to laterals, clusters of new rootlets occurred. These originated in clumps from the enlarged, bulging, meristematic areas, usually 5 to 25 occurring in a single cluster. From 500 to 1,500 were counted on individual plants. The roots were hairlike, a few millimeters to 7 inches in length and invariably pursued a horizontal course in the mellow, moist soil. Thus the older surface-absorbing system, only fragments of which remained intact, was being replaced by this newer one. The deeper and younger portions of the old root system were apparently functioning in a normal manner. Further studies were made at the end of February. The plants had intermittently grown during the warmer winter weather, since soil moisture was very favourable. The new roots, formed in the fall, were now 8 to 16 inches long. They had not deviated from their horizontal course. The larger ones were well furnished with numerous short laterals which greatly increased their absorbing area. Moreover, new laterals were developing on the old taproot to a depth of 2 feet as well as on the larger laterals near their place of origin.

MATURE PLANTS

The plants had reached their maximum development. The numerous leafy stalks, 15 to 40 in number and 8 to 16 millimeters in diameter, reached heights of 30 to 42 inches. They were just ready to blossom. Roots were exceedingly abundant in the surface foot of soil; on one plant there were 109 with a diameter of 1 to 2 millimeters and 430 finer ones in this soil layer. A maximum lateral spread of 30 inches was found. For example, one strong lateral, originating at a depth of 4 inches, ran obliquely outward and downward to a depth of 29 inches at a horizontal distance of 27 inches from the plant. Here it turned downward and ended 1 foot deeper. This illustrates the marked development of some of the new roots. Most of them, however, were shallower and shorter. The horizontal course of those arising within 2 to 3 inches of the surface and the more oblique direction of growth of those originating somewhat deeper were very characteristic. With its dense network of branches, this new root system rather thoroughly occupied the surface 12 inches as well as portions of the deeper soil. It greatly supplemented the activities of the deeper portion of the old root system.

The taproots reached a maximum depth of 55 inches. Strong, much-branched laterals occurred mostly in the second foot of soil. These spread 12 to 18 inches from the taproot and then often turned downward. Throughout its course below the first foot short laterals arose from the taproot at the rate of 4 to 12 per inch.

SUMMARY

Carrots are characterized by a strong, deep, well-developed taproot system. Plants with tops in the fifth-leaf stage have taproots 2.5 feet deep. Branching throughout is very poor. The greatest branching is in the surface 2 to 4 inches of soil where a few laterals extend horizontally 8 to 10 inches. During the following month the taproot grows 2 feet deeper. A fairly profuse network of horizontal branches extends from the lower half of the fleshy portion of the taproot 1.5 to 2 feet outward into the surface soil.

A few major branches descend rather vertically, supplementing the absorbing area of the now better-branched taproot, to a depth of nearly 3 feet. About the middle of August, maturing, plants have well-formed "carrots" from which fine roots arise in great abundance. These furnish an excellent surface-absorbing system near the plant. Many of the formerly horizontal laterals have turned vertically downward after reaching a maximum spread of 2.3 feet. They give rise to small laterals only but extend through the third and fourth and often the fifth foot of soil. Well-branched taproots reach the 7.5-foot level. The roots are still growing vigorously and undoubtedly extend much deeper.

When roots die from drought, they are replaced by multitudes of new ones of a very fibrous nature but often 1.5 to 2.5 feet in length. A dry surface soil tends to promote a vigorous development of strong laterals from the deeper portions of the taproot.

OTHER INVESTIGATIONS ON CARROTS

Carrots of the French Forcing and Long Red Altrincham varieties were examined in the middle of September at Geneva, N. Y. On both the taproot was small and soon tapered into a filament. We traced it downward 16 inches, at which depth it was too delicate to follow further. The horizontal roots apparently extended little more than 1 foot.

The fibrous roots chiefly proceeded from the taproot, though a few started from near the base of the thickened part.

These extended both deeply and shallowly, some rising nearly or quite to the surface, while others apparently penetrated as deeply as did the taproot. It seems clear in this case that the fineness of the root was an obstacle to its complete recovery by the method employed of washing away the soil. Carrots of the Long Orange variety were set out in the spring at the same station and examined late in September of the second year of growth.

The leading roots were found to extend quite as far as those in plants of the first year's growth. The fibrous roots, however, appeared less numerous. No branches from the horizontal roots extended upward to near the soil surface as was the case in both of the varieties grown for a single season. The lower roots were developed to a greater extent than in those plants grown directly from seed.

At Saratov in southeastern Russia, the roots of the carrot have been traced, to a depth of 40 to 52 inches. Branching was very abundant in the plowed soil layer and the plant thus well adapted to absorb the water afforded by the summer precipitation.

At greater depths the branches were larger and occupied the soil rather thoroughly, being especially well developed where they occupied old burrows and earthworm holes. Investigators in Germany state that the roots of carrots are characterized by their great depth of penetration which was found to be about 5 feet. The root system was much like that of the lettuce in so far as the separate root branches penetrated rather vertically downward and did not spread widely. But they differed in that adventitious roots were lacking in the carrot and branches of the first order furnished the means of lateral spread. These laterals were found to be rebranched in a manner similar to the lettuce.

Experiments in the vegetable gardens at Ithaca, N. Y., where carrots were grown in a gravelly sandy loam are of interest. The plants were grown in rows 18 inches apart. The taproot and several other larger roots arising from the side of the carrot reached depths of 30 inches. These roots produced almost countless numbers of branches which were themselves rebranched at least to the second order. The soil directly beneath the plant was filled with roots to a depth of 25 to 30 inches. A space 4 to 6 inches wide in the center between the rows was not so completely filled although at a depth of 4 to 8 inches many roots met and overlapped between the rows.

ROOT HABIT IN RELATION TO CULTURAL PRACTICE

The amount of soil that must be pushed aside and the extensive development of roots in the surface soil helps to make clear the benefits of a deep, loose, mellow soil. Such a soil is also beneficial in preventing the formation of a soil crust over the slowly germinating seeds and about the slowly developing plants.

The carrot has a delicate root system during its early stages of growth and one poorly adapted to penetrate stiff, hard soil. Muck, a fine, loose-textured soil, is an ideal one for the growth of carrots and indeed for most root crops. Since the plants grow slowly, they cannot successfully compete with weeds; hence the value of clean and thorough but shallow cultivation. It has been shown at Ithaca, N. Y., using both early and late carrot crops, that little or no benefit is derived from cultivation other than preventing weed growth.

Owing to the shading of the soil by the tops and the consequent lowering of soil surface evaporation, together with the thorough occupancy of the shallow soil by a dense root growth, soil moisture was not conserved by cultivation. Plats where the weeds were kept out by scraping the surface usually had slightly less moisture than those repeatedly cultivated, but in some cases these conditions were reversed.

For the best growth of the plant the usual spacing, 4 to 8 inches apart in rows 12 to 18 inches distant, is too close. Competition between closely spaced plants results in considerable dwarfing. When vegetable crops are crowded into a small area, garden practices, such as furnishing the plants with unusually large supplies of nutrients, conserving the water supply, and eliminating weeds, are employed to stimulate growth. Competition, however, is always severest where plants of the same kind are grown in close proximity. All make the same demands for water, nutrients, and light at the same levels and at the same time. The vegetable grower is concerned, however, with the best interests of the plant only in so far as they meet his needs. By growing carrots thickly in favourably rich, moist, loose soil, he prevents the plants from attaining too great a size and secures an abundance of well-formed, succulent, medium-sized roots.

Index

D

E

F

G

H

I

R

S

T

U

V

W

Z